SCARECROW PROFESSIONAL INTELLIGENCE EDUCATION SERIES
Series Editor: Jan Goldman

In this post–September 11, 2001 era, there has been rapid growth in the number of professional intelligence training and educational programs across the United States and abroad. Colleges and universities, as well as high schools, are developing programs and courses in homeland security, intelligence analysis, and law enforcement, in support of national security.

The Scarecrow Professional Intelligence Education Series (SPIES) was first designed for individuals studying for careers in intelligence and to help improve the skills of those already in the profession; however, it was also developed to educate the public in how intelligence work is conducted and should be conducted in this important and vital profession.

1. *Communicating with Intelligence: Writing and Briefing in the Intelligence and National Security Communities*, by James S. Major. 2008.
2. *A Spy's Résumé: Confessions of a Maverick Intelligence Professional and Misadventure Capitalist*, by Marc Anthony Viola. 2008.
3. *An Introduction to Intelligence Research and Analysis*, by Jerome Clauser, revised and edited by Jan Goldman. 2008.
4. *Writing Classified and Unclassified Papers for National Security: A Scarecrow Professional Intelligence Educational Series Manual*, by James S. Major. 2009.
5. *Strategic Intelligence: A Handbook for Practitioners, Managers, and Users*, revised edition by Don McDowell. 2009.
6. *Partly Cloudy: Ethics in War, Espionage, Covert Action, and Interrogation*, by David L. Perry. 2009.
7. *Tokyo Rose / An American Patriot: A Dual Biography*, by Frederick P. Close. 2010.
8. *Ethics of Spying: A Reader for the Intelligence Professional*, edited by Jan Goldman. 2006.
9. *Ethics of Spying: A Reader for the Intelligence Professional*, Volume 2, edited by Jan Goldman. 2010.
10. *A Woman's War: The Professional and Personal Journey of the Navy's First African American Female Intelligence Officer*, by Gail Harris, 2010.
11. *Handbook of Scientific Methods of Inquiry for Intelligence Analysis*, by Hank Prunckun, 2010.

D0781968

Handbook of Scientific Methods of Inquiry for Intelligence Analysis

Hank Prunckun

Jan Goldman
Series Editor

*Scarecrow Professional Intelligence
Education Series, No. 11*

THE SCARECROW PRESS, INC.
Lanham • Toronto • Plymouth, UK
2010

Published by Scarecrow Press, Inc.
A wholly owned subsidiary of The Rowman & Littlefield Publishing Group, Inc.
4501 Forbes Boulevard, Suite 200, Lanham, Maryland 20706
http://www.scarecrowpress.com

Estover Road, Plymouth PL6 7PY, United Kingdom

British Library Cataloguing in Publication Information Available

Library of Congress Cataloging-in-Publication Data

Prunckun, Hank.
  Handbook of scientific methods of inquiry for intelligence analysis / Hank
Prunckun.
     p. cm. — (Scarecrow professional intelligence education series ; no. 11)
  Includes bibliographical references.
  ISBN 978-0-8108-7191-5 (hardcover : alk. paper) — ISBN 978-0-8108-6753-6
(pbk. : alk. paper) — ISBN 978-0-8108-7381-0 (eBook)
  1. Intelligence service—Methodology—Handbooks, manuals, etc. 2. Science—
Methodology—Handbooks, manuals, etc. 3. Social sciences—Methodology—
Handbooks, manuals, etc. 4. Behavioral assessment—Methodology—
Handbooks, manuals, etc. I. Title.
JF1525.I6P78 2010
327.12072—dc22   2009038488

∞™ The paper used in this publication meets the minimum requirements of
American National Standard for Information Sciences—Permanence of Paper for
Printed Library Materials, ANSI/NISO Z39.48-1992.

Printed in the United States of America

To Ann, for all you have done.

Contents

Editor's Foreword

The goal of this series is first and foremost to develop and contribute to the educational literature of the intelligence profession. While other intelligence book series discuss "what to do with intelligence" by focusing on issues and policy, the goal of this series will always be "how to do intelligence." Previous books in the series have dealt with methodologies of intelligence analysis. Although there is no agreed-upon method of how to process raw intelligence into a finished product, this book contributes to a long line of theories and viewpoints on the subject. Hank Prunckun is the second Australian intelligence expert to contribute to this series. The first book, and extremely well-received by the intelligence community, was written by Don McDowell. Both books highlight the worldwide interest of improving intelligence analysts beyond any country's borders. It is hoped that as this series develops, additional analytical books will be added, allowing intelligence professionals to decide for themselves which method works best.

Jan Goldman
Series Editor

Preface

It has been argued that since the terrorist attacks of September 11, 2001, no other profession has experienced change to the same extent as that of intelligence. It has grown much larger and its mission is more complex. As an indicator, government and private-sector security agencies have recruited intelligence analysts to process what has become a voluminous amount of raw information flowing into these agencies' data-collection systems.

Unfortunately, there is an unmet need for analysts who are able to process these data. For this reason, a growing number of colleges and universities offer intelligence training, so candidates for analyst positions can begin their duties without protracted on-the-job instruction. This book offers students in such courses a way to gain the analytic skills essential in undertaking intelligence work. It is therefore a handbook for analysts and other scholar spies.

This book addresses a subject that is inadequately covered in the current body of intelligence literature. Although there are many research texts available, they focus mainly on social- or behavioral-science research. Despite these academic disciplines being allied to intelligence research, a comprehensive text covering the topic from the point of view of *secret intelligence* is absent.

While the literature on intelligence abounds with works on spy gadgetry and covert surveillance, one has to look far and wide for anything on intelligence research and analysis. This book aims to acquaint the reader with how intelligence fits into the larger research framework. It covers not only the essentials of applied research but also explains the function, structure, and operational methods involved in intelligence work. In particular, it looks at how an analyst works with classified information in a security-conscious environment and obtains data via covert methods. It also looks at how intelligence data are validated, in marked contrast to how a social-science researcher performs the same task. The reader is left with little doubt about what intelligence is, how intelligence is developed, and how it is processed for a profession that has security and secrecy at its core.

The need for such a book was borne out of my personal experience as an analyst. In many of the positions I held during my career, I relied on texts in other academic disciplines as none addressed the craft of intelligence. Occasionally I found texts in the field of criminal justice and criminology that were of value, but again they addressed issues faced by analysts obliquely. There are several excellent texts for law-enforcement intelligence, but by definition they are narrow in focus (e.g., police-centric) and do not apply the principles of intelligence holistically.

Handbook of Scientific Methods of Inquiry for Intelligence Analysis examines how social- and behavioral-science research methods can be applied to intelligence work. It is a systematic presentation of the concepts within the intelligence discipline, thereby providing students with practical knowledge of how to be effective researchers in a variety of intelligence settings: military, national security, law enforcement, business, and the private sector.

The book comprises fifteen topics that can be studied over one semester. Each topic contains a number of concepts that build into a thorough understanding of intelligence research and analysis. At the end of each chapter is a list of key words and phrases, a number of study questions, and learning activities for students to test their understanding of intelligence research.

ACKNOWLEDGMENTS

I'd like to thank several people for making this book possible: the series editor, Jan Goldman, for his interest in my work; Victoria Herrington, my colleague at the Australian Graduate School of Policing, Charles Stuart University, for her comments on counterterrorism; and finally my postgraduate students, who provided feedback from a scholar's point of view—Luke Langtry, Vicki Rogers, Natalie Carroll, Nathan McGrath, Paul Robinson, and especially Sarah Eastlake-Smith, whose editorial suggestions added much to improving the final presentation. Nevertheless, any shortcomings remain mine and mine alone.

I'd also like to extend a number of belated thanks. To Big Mike, Little Mike, Evan, and Pudgy, who many years ago shared an interest, though only passing, for the art of spying; it was their interest that fueled my passion for the profession in which I have since worked. To Wally and my Uncle Ed, for the interest they shared with me in radio and electronics, which eventually led to my involvement in radio engineering and the application of this knowledge to secret intelligence. To Evan, my cousin Ted, and my father, who showed me the subtleties involved in covert photography. And to my father, who, as a former private detetective and auxilliary police officer, started me off on my career as a scholar-spy. Finally, to my wife, Ann Doolette, for her encouragement and counsel throughout the project—I dedicate this book to you.

1

🌀

The Fundamentals of Intelligence

This topic provides an introduction to intelligence research by examining:

1. Intelligence versus information;
2. Intelligence defined;
3. Intelligence as knowledge;
4. Intelligence as a process;
5. Taxonomy of intelligence research;
6. Anatomy of intelligence;
7. Typology of intelligence; and
8. Intelligence: a quick glance at its history.

INTRODUCTION

Why be concerned with intelligence? Because intelligence enables one to exercise control over a given situation. In this sense, *control* equates to *power*. Ira Cohen, in his classic treatment of the study of power, wrote:

Power is sought because without power the security and even the ability of [one] to continue to exist is generally decreased. Without power, [one] has no ability to deter another . . . from actions whose consequences threaten the vital interests of the former. Without power [one] cannot cause another . . . to do that which the former desires but which the latter desires not to do. Power is sought because the more power that [one] has, the greater is the number of [his or her] available options. The more options available to [one], the greater

1

[his or her] security. The greater [his or her] security, the better off [he or she is]. [He or she is] more secure in [his or her] life and in the enjoyment of [his or her] private property.[1]

Intelligence is, therefore, not a form of clairvoyance used to predict the future but an exact science based on sound quantitative and qualitative research methods. Intelligence enables the analyst to present solutions or options to decision makers based on defensible conclusions. At this juncture, it should be noted that such conclusions are not absolute, and there will always be some level of probability or uncertainty involved with presenting intelligence findings. Nevertheless, uncertainty can be reduced and conclusion limits further defined so decision makers understand the boundaries. This must be contrasted with making decisions based on "gut feel," "belief," "faith," or "trust."

Having said that, the word *intelligence* conjures up assorted notions of spying and espionage, secrets, and the world of exotic gadgetry. Yet to others, the word *intelligence* is closely associated with the Orwellian concept of "big brother"—a world of hardball politics and an uncompromising quest for influence.

To some degree, intelligence work is associated with these concepts, but here the study of intelligence is approached from the focus of the analytic methods that turn information into intelligence. This process is based on methods used in applied research rather than the James Bond-like devices used by cinema heroes or in the authoritarian oppression exercised by police states.

In the post–September 11, 2001, world, colleges and universities across the globe have responded to the need to develop intelligence courses for the new cadre of analysts needed to support national security. Much of what is taught in these courses will also be applicable to other types of intelligence: law enforcement, military, business, and private intelligence. The growth of these educational programs means that training aids are also needed to instruct new analysts in the scientific methods of inquiry for intelligence research.

Information is the unrefined raw material used to produce finished, focused *intelligence*. Without information, intelligence could not exist.

INFORMATION VERSUS INTELLIGENCE

Trying to define information is difficult but not impossible. Information is like gravity and electricity, as it cannot be defined by tangible examples.

Nevertheless, its properties can be observed and described, thus enabling improvement in the analytic methods that produce intelligence. The problem hard sciences face in trying to define gravity and electricity has never prevented engineers from designing and building applications that involve these phenomena. Therefore, a lack of a physical variable does not prevent analysts from producing intelligence from what we call *information*.

It is quite safe to say that every facet of our lives, whether central or incidental, is in some way related to information. We rely on an alarm clock to wake us in the morning, the newspaper to tell us what is happening in the world beyond the end of our street, the radio to alert us if rain is expected, an array of indicator lights and meters on our car's dashboard to tell us about the car's performance as we drive to work, traffic lights and signs to alert us to road conditions, and on we could go until the clock tells us it's time to lay our work aside and to go off to sleep.

Individuals, organizations, and indeed whole societies owe their survival to information. The concept of community is only possible because of our ability to collect, store, retrieve, and transfer information from one person or body corporate to another. The more complex our society, the more it necessitates the conversion of information into intelligence.

> The late Colonel Russell J. Bowen, a U.S. army and CIA intelligence analyst is attributed with saying: "Religion and intelligence are two sides of the same coin: both are institutionalizations of man's attempts to cope with his fear of the unknown; one in the spiritual realm, the other in the practical."

INTELLIGENCE DEFINED

The term *intelligence* has four meanings:

1. Actions or processes used to produce knowledge;
2. The body of knowledge thereby produced;[2]
3. Organizations that deal in knowledge (e.g., an intelligence agency); and
4. The reports and briefings produced in the process or by such organizations.[3]

In this book, intelligence as a process is categorized by the different functions it performs. *Knowledge* in the context of intelligence equates to *insight*, or viewed another way, the ability to *reduce uncertainty*. Insight and certainty offer mankind the ability to make decisions that enable civilizations to take better control over the "unknown."

INTELLIGENCE AS KNOWLEDGE

As a body of knowledge, intelligence deals with an adversary, a potential adversary, or a possible area of operation that is useful to managers in planning and carrying out their organization's mandate. Terms like *target, subject, person of interest, subject of interest* are some of the ways intelligence manifests itself as knowledge. To demonstrate, consider the following notional examples:

Law Enforcement Context. We have intelligence indicating Mack Da-Knife is planning to break into the Pine Point office of the Springfield Credit Union this Friday night.

Business Context. Intelligence suggests Nerro Entertainment is about to begin an advertising campaign in the northeast this autumn, attempting to capture customers in the 21- to 41-year-old range.

Military Context. We have recently received intelligence indicating the French government has authorized a nuclear weapons test. This intelligence indicates it will take place at their Pacific test site at Mururoa Atoll during the week beginning July 16.

National Security Context. Intelligence from agents in the Caribbean alerts us to the imminent passage of legislation by Cuba which will legalize a multiparty, democratic political system.

INTELLIGENCE AS A PROCESS

The intelligence process is a series of procedures or steps, forming the *intelligence cycle*. The cycle is initiated by a decision maker who poses a question or requests advice. This is termed an *intelligence requirement* (in some intelligence agencies, such as the military, this is referred to as *essential elements of intelligence*—EEI). The intelligence requirements are forwarded to an intelligence agency and the cycle begins.

The intelligence cycle (see fig. 1.1) consists of seven steps, with the first five converting the raw data into finished, focused intelligence:

1. Direction setting (i.e., problem formulation and planning);
2. Information collection;
3. Data collation;
4. Data manipulation and processing; and
5. Data analysis.

This resulting intelligence is then treated with two further steps:

1. Report writing; and
2. Dissemination to decision makers (which would include provision for feedback).

Depending upon the initial intelligence requirements (e.g., the operational objective), a single "loop" maybe sufficient to complete the intelligence research project and provide the decision maker with the insight sought. However, in practice, further data may need to be collected with the cycle beginning again, or the cycle may have two or more tasks being performed at once and may double back before advancing again. For instance, once the research question has been formulated and the data collection plan devised, an outline of the report may begin, and as the more readily available pieces of information flow in, a database or spreadsheet may have been constructed and the data collated.

Furthermore, even before all the data are received, some preliminary analysis may be carried out, and depending upon the results, further information may be requested (e.g., if by chance these results show the data would be inadequate to answer the research question or a serious limitation

Figure 1.1. The intelligence cycle.

noted). This would mean that the data collection plan is revised and field operatives called on to gather more or different data, and so on.

As long as a specific intelligence operation is being conducted, the analytic process will be continuous—forming a cycle. As new information is being collected, other data will be stored and analyzed. The resulting outcomes will be disseminated for either immediate use and/or used to set new collection goals.

The dissemination of the intelligence product can take a variety of forms. Take for instance the case of business intelligence—it could be a background history on a company or one of its executives, a diagram of a company's office layout, identification of new projects being researched, a prediction on the intended release of a new product, staff salaries, the classification and number of personnel on a company's payroll, and so on.

The intelligence cycle is not unique to intelligence research but has parallels with research cycles in other academic disciplines.[4] For instance, the research cycle that is used in applied social research shares the same pattern:

- Establish a plan for information collection and carry out initial field work;
- Observe, discuss and collect data;
- Analyze the data and write the report; and
- Distribute the report and gather feedback that can be used to formulate further disseminate strategies.

TAXONOMY OF INTELLIGENCE RESEARCH

Intelligence is classified into three categories: tactical, strategic, and operational. Tactical intelligence is information that contributes directly to the achievement of an immediate goal, whereas strategic intelligence relates to long-term forecasts or broader conclusions on larger objectives. Operational intelligence provides support to an operation that is either underway or about to begin.

Tactical Intelligence

- Is short-range or time limited; and
- Consists of patterns or operational mode activities.

Strategic Intelligence

- Considered to be a higher form of intelligence;
- Provides a comprehensive view of a target or an activity;

- Comments on future possibilities or identifies potential issues;
- Provides advice on threats, risks, and vulnerabilities;
- Provides options for planning and policy development;
- Assists in allocating resources; and
- Requires extensive knowledge of the target or the area of activity.

Operational Intelligence

- Provides immediate insight that supports an operation; and
- Oriented towards a specific target or an activity.

Although there appears to be an obvious separation between these classes of intelligence, in certain situations a given piece of information may be relevant to more than one—say, to meet a tactical objective as well as a strategic goal.

In military intelligence, there are many categories of strategic, tactical, and operational intelligence specific to a branch of an armed service, such as *combat intelligence* to the army:

- Provides military commanders with advice on the threat posed by an enemy through a process known as *intelligence preparation of the battlefield* (IPB);
- Provides knowledge of an enemy's *order of battle*—that is, a list of military units, the type of equipment it carries, and the capabilities of that equipment, as well as the location of the units and other information specific to the battlefield environment;
- Provides analysis of the weather and geographical features likely to be encountered by a commander when conducting combat operations; and
- Assists commanders in executing existing plans that are based on sound decisions—these decisions take into account the enemy's intentions, capabilities, vulnerabilities, and, therefore, the likely courses of action.

Naval intelligence has categories that are specific to its mission, such as intelligence for amphibious operations, intelligence for antisubmarine warfare, and intelligence for air operations. The air force has categories within this taxonomy of strategic, tactical, and operational intelligence applicable to its areas of concern and operations in air and space as well as information warfare in cyber space—for instance, indications and warning intelligence and target intelligence (i.e., target development and battle damage assessment).

ANATOMY OF INTELLIGENCE

Just as the human anatomy is comprised of different components, intelligence is also comprised of components: applied intelligence research, counterintelligence, espionage, counterespionage, and covert action.

Applied Intelligence Research

Basic research, or theoretical research as it is sometimes known in other academic disciplines, is concerned with research for its own sake—that is, when undertaken, it has no practical application in mind. It is knowledge for knowledge's sake. The findings of such research are sometimes used later in an applied setting, but at the time of conducting the research, this was not the aim.

In contrast, applied research has a practical purpose—to offer a basis for making a decision (i.e., to provide insight or reduce uncertainty). Intelligence is in this sense applied research—it is the outcome of processing raw information that has been collected from a variety of sources—open, semi-open, official, clandestine, or covert. Once the information is in the hands of an intelligence analyst, it is evaluated and any irrelevant information discarded. The pieces of information pertinent to the matter under investigation are then analyzed, interpreted, and formed into a finished "product."

This product can take the form of an oral briefing, a written briefing, a target profile, a tactical assessment, a strategic estimate, or any number of other forms of reports. These products are them disseminated to the end user (the customer). The intelligence process can be summarized as analysis that leads to the production of deep, thorough, or meaningful understanding about a particular matter.

Counterintelligence

Counterintelligence is concerned with neutralizing or destroying the effectiveness of an adversary's intelligence activities, which makes it a security function. The thrust of counterintelligence is to protect an organization from infiltration by an adversary, to protect against inadvertent leakage of confidential information, and to secure its installations and material against theft and sabotage.

As with tactical and strategic intelligence, there is a thin line dividing applied intelligence research and counterintelligence. That is, information concerning an adversary's attempts to penetrate one's organization can feed into the intelligence cycle, revealing opponents' information voids as well as their capabilities and possible intentions. Chapter 15 focuses more specifically on some of these counterintelligence issues.

Espionage

> Espionage is not a James Bond-like game. Rather, it is a serious business that can have deadly consequences. The Wall of Honor at CIA headquarters holds testimony to this fact. In 2002, it was reported to hold 79 stars, each representing an officer of the agency who gave his or her life in the service of country. Forty-eight of those officers have their names listed in the *Book of Honor*,[5] but the remaining (at that time) were anonymous as their services to their nation were still classified.[6]

This is the classic form of information gathering dating back centuries, and it forms part of the second step of the intelligence cycle. Espionage, or simply spying, traditionally utilizes undercover agents. A distinction must be made that an *agent* is someone who acts on behalf of another person or agency (e.g., FBI agent) whereas an *officer* (or an *operative*) is someone who is charged with an authority that requires them to discharge a statutory responsibility (e.g., Boston police officer). An agent is someone whom an intelligence officer recruits to obtain secrets on behalf of the operative's government.[7] In such situations, the recruiting officer is termed a *case officer*.

These agents (undercover) are placed in positions allowing them to view, overhear, or otherwise obtain information that could not be gained in any other way. However, with technological improvements, more technical means of espionage (i.e., technical surveillance and unobtrusive methods) are favored over the classic use of espionage agents. It could be said that this worked reasonably well when intelligence agencies faced actors that were states.

But since the al-Qaeda terrorist attacks of September 11, 2001, intelligence agencies recognized the importance of having agents in place to gather data. One reason for the shift to technically gathered data was the comparatively high cost of running field agents and improving the reliability of the data collected (e.g., aerial and satellite photographs are not susceptible to exaggerating the truth as an agent might be. These data simply show what is there and what is not).

The events of September 11 and the subsequent terrorist attacks in Madrid, London, and Bali show that the advantages of technically gathered data were of little value against terrorist cells operating in a vastly different fashion from those of, say, a foreign government's military. This type of confrontation, and other nontraditional challenges to a state-centric paradigm, no longer applies. Nations now face threats from weak and corrupt governments, rogue states, sub-state and trans-state actors, as well as international, organized criminal groups, radical ethnic and religious groups, and right-wing political groups. All of these threats pose special data collection problems that defy a purely technical approach.

When it comes to describing *cover*—a plausible story about all facets of the operative's life—there are essentially two types: official cover and nonofficial cover (NOC, pronounced *knock*). NOC is also referred to as commercial cover when the operative works for a phantom company created and maintained by an intelligence agency. The former are personnel posing as government employees of some description, and the latter are those who on the surface have no connection with government.

The NOC operatives have been described as the truest practitioners of espionage as they operate on their own at all times with no protection from their government. In the case of foreign espionage, if they are caught abroad, they may be tortured during interrogation and perhaps executed. If this happens, no media conference will be held, and no one will hear about the event. NOC operatives operate alone and die alone.

Nonetheless, espionage still employs audio surveillance devices, radio frequency devices, and special photographic equipment, including space-based reconnaissance satellites. The use of such tools can provide the intelligence analyst with an exponential gain in both the quantity and, under the right circumstances, the quality of the information gathered. However, it is at the peril of the intelligence agency that it neglects data collection by human sources. Because the espionage function features so heavily in intelligence work, chapter 6 explores various forms of covert information gathering.

Counterespionage

"Counterespionage is often touted as the aristocratic sector of secret operations. In the romantic image the counterespionage man is pitted against his fellow professionals on the other side who are trying to get his nation's secrets. His job is to foil them. It is a true adversary relationship unlike the espionage situation, in which two men work together to purloin secrets. Most spy stories are not about spying but about counterspying."[8]

On the surface, counterespionage presents as being simple spying but is in some ways related to counterintelligence. It is a precise function that is the most subtle and sophisticated of all intelligence functions. It calls for the engineering of complex strategies that deliberately put one's agent(s) in contact with an adversary's intelligence personnel. This is done so that an adversary can be fed with disinformation which will hopefully lead to confusion, thus disrupting the adversary and allowing the perpetrator to prosper.

Covert Action

This intelligence function lies in a somewhat gray area of intelligence work. It uses the various methods of information gathering and analysis but incorporates areas including advice and counsel, financial and material support, as well as technical assistance to individuals, groups, or businesses that are opposed to, or working in competition with, a target or adversary.

Covert action is a function by which the perpetrator uses the information it collects to strengthen its allies and to weaken or destroy its opponents.[9] The effectiveness of covert action is contingent upon the perpetrator's involvement remaining hidden, or at the very least, deniable.

On the one hand, if a "plausible denial" can be maintained, then the rewards of such ventures can be enormous. On the other hand, if the perpetrator's involvement is discovered, the consequences of this activity can be catastrophic (e.g., the French government's involvement in the sabotage of the Greenpeace ship, *Rainbow Warrior*, in Auckland Harbour, New Zealand on July 10, 1985, by intelligence operatives of the *Direction Générale de la Sécurité Extérieure* [DGSE] or in English, the Directorate-General for External Security).

Figure 1.2.   Rainbow Warrior II docked at Port Adelaide, South Australia.

TYPOLOGY OF INTELLIGENCE

Intelligence is structured according to type, and the typology is based on the environment in which the organization operates. There are five major types of intelligence, including national security (which includes foreign policy and international politics), military, law enforcement, business, and private. A sixth type, emergency services (e.g., firefighters and search and rescue teams), also has intelligence cells, but although they are entitled *intelligence,* they do not perform the same function or perform the same level of analysis as those discussed here.[10, 11] Because of these limitations, this intelligence type is not considered within the pages of this book.

It is important to note these environments can overlap—for example, an investigation into the capability of a terrorist cell may be of interest to local law enforcement agencies as well as to agencies involved in national security, the military, and some private security firms. Moreover, with regard to military intelligence, it is in some cases intimately aligned with national security because it not only informs military commanders of the intent and capabilities of an adversary but also political leaders who are responsible for authorizing the use of military force and directing strategic military policy.

In addition to the overlap or close working partnership, the same methods of operation, tactics, devices, information storage systems, and methods of analysis are used by each intelligence type. This is because information holds no bounds as to its usefulness, and a particular piece of data could conceivably be the target for more than one type of intelligence user. In other words, the primary difference between the various types of intelligence lies in the end use or general thrust of the intelligence operation.

National Security Intelligence

National security intelligence is conducted by the various branches of a nation's armed forces, foreign diplomatic service, and, depending on the country, its atomic energy authority. It is sometimes referred to as *foreign policy* intelligence, depending on the context. Western nations generally tend to have a central agency that acts to coordinate subsidiary intelligence agencies and the collection and processing of information from all sources. Other nations, in contrast, have a unified system with one supreme agency taking on all three roles—coordination, collection, and analysis.

The types of information sought by national security intelligence analysts can be anything from the current political issues facing a foreign government; the health, education, and social structures of the country; its social problems; and its legal institutions. They may include issues concerning the availability of food production and distribution, world resources (e.g., oil and potable water), international trade relationships, world migration patterns and changes in the ethnic composition of nations, as well as the state of the global monetary order. Without a doubt, they seek also information on foreign technological developments, nuclear matters, and almost anything to do with foreign weapons production, defense industries, defense installations, and military capabilities.

Military Intelligence

Military decisions carry a heavy burden of responsibility. As such, these decisions not only impact the lives of the fighting forces but also a nation's liberty. Military intelligence concerns itself with matters key to fighting a war: "enemy strength, capabilities, and vulnerabilities as well as information on weather and terrain."[12] In addition to dealing with these fundamental concerns, intelligence produced by the military "has to be timely, accurate, adequate, and usable."[13]

Military intelligence is decision making by commanders with regard to either the operational environment, forces (whether they are hostile, friendly, or neutral), as well as the civilian population in the operational (or potential) area. A nation's military will carry out intelligence activities regardless of whether it is at war or at peace (i.e., to prevent a surprise attack or to transition to a war footing at short notice) and at the three levels—strategic, tactical, and operational.

It would be most unusual for any army, navy, or air force not to have some form of a military intelligence capability. It may take the form of a specialist unit, or it may be part of another government service. Staff can be from the military or civilians assigned to the intelligence agency because of their particular technical or analytical abilities. Military personnel who do not have such skills are often trained at special colleges, which are set up specifically for this purpose.

Law Enforcement Intelligence

Law enforcement intelligence aims to increase the accuracy of decisions made by commanders. Intelligence provides senior officers with advice needed to make sound decisions and in this regard provides a focus on those criminal activities that would generally go undetected until they evolve into a community problem.

These agencies are much wider than just police and include compliance and regulatory agencies that perform law enforcement functions (which can be quite numerous), such as immigration, customs services, and prison intelligence units. Law enforcement intelligence also encompasses agencies engaged in combating the threat from foreign and internal subversion, espionage, sabotage, and terrorism. Other agencies protect national security and foreign policy interests by enforcing export regulations relating to prohibited dual-use items, such as certain hardware technology, software, chemicals, and nuclear material (e.g., U.S. Department of Commerce's Office of Export Enforcement).

Business Intelligence

"The business world is an arena of competition. Like hustling baseball clubs, campaigning politicians, and battling armies, companies are in conflict with their counterparts."[14]

Business intelligence is concerned with the acquisition of trade-related secrets and commercial information that is held confidential from competing firms. Although the media has exposed much about the unethical behavior of some intelligence practitioners, it is safe to say that a good deal of information is gathered through open- and semi-open sources. The focus of this activity can be on several levels—local, regional, national, or even international. Business intelligence is not only limited to the realms of corporations and businesses themselves but also can include private investigation firms that specialize in this area and intelligence agencies of foreign nations. The former often comprises spies-for-hire and the latter foreign military and national security agencies that are targeting trade secrets.

Private Intelligence

The structure of private intelligence is very diverse, but for the purpose of this book, it will be limited to those firms and private agents who offer their services in intelligence work for fee or reward to the public. Although the term *private* implies an individual, there is some overlap in what constitutes private intelligence and what may be business intelligence or even national security intelligence. The ultimate determiner is who is contracting the "spy-for-hire."

Private intelligence practitioners offer a range of specialist services that go beyond the bounds of the average private investigator or private detective. Often, the private intelligence practitioner comes from a background in law enforcement, military, or national security intelligence work.

Their specialties may be in background investigations or surveillance. They may have extensive training in the use of state-of-the-art optical or electronic audio surveillance equipment, and they would be familiar with the techniques of intelligence analysis. They may offer advice on business counterintelligence and electronic audio countermeasures (debugging). They may also specialize in providing close personal protection for important public figures, crisis management, or business continuity planning.

Private intelligence agencies could (arguably) include commercial organizations that maintain databases for specialized inquiry work, for example, credit reporting. Likewise, private intelligence agencies might even encompass what are called *policy institutes* (or think tanks) where research agencies engage in scholarly investigation for fee paying clients. Private intelligence practitioners are being viewed by some of their government counterparts as a viable supplementary alternative deemed necessary in cases where resources are constrained.

INTELLIGENCE: A QUICK GLANCE AT ITS HISTORY

The operational aspects of military intelligence are remarkably similar to those of law enforcement and business intelligence. It is not surprising then to find that these forms of intelligence find their ancestry in this genetic stock.

A cursory examination will suffice in demonstrating the lineage between these intelligence relatives. Such a comparison acts to reinforce the underlining intelligence theory—the theory developed and refined by the military and adopted by its "offspring."

The history of military intelligence dates back many centuries, and isolated examples of the craft can be cited in events from biblical times and earlier. However, it was not until the last 150 years or so that military intelligence came of age. Its official birth was registered with the formal creation of intelligence divisions within various countries' war departments.

Between 1600 and until just after the Second World War, European nations began developing extensive intelligence systems but, compared to today, without great success. This is evidenced by France's miscalculation of the size of the German army at just half of what it really was at the outbreak of the First World War.

Between the First World War and the Second World War, military intelligence services expanded greatly in scope and sophistication. The only exception was that of America, whose intelligence system was to some extent disassembled. In 1929, the then U.S. Secretary of State, Henry L. Stimson, advanced the now infamous dictum that "gentlemen do not read each other's mail."[15]

Stimson's comments were in response to having learned the existence of Herbert O. Yardley's "Black Chamber." Stimson is reported to have rejected any argument that justified covert code-breaking operations. He strongly disapproved of Yardley's secret activities, regarding it a low, dirty business that violated the principle of mutual trust upon which, in Stimson's view, foreign policy should be based. Stimson then shut down Yardley's code-breaking operations. History has shown the fate America suffered in the years leading up to the Second World War because of Stimson's decision to restrict intelligence to decision makers.

World War II dramatically changed any misconceptions political leaders had about the role intelligence could play; its importance to military planning and operations today is unquestioned. The intelligence offspring of military intelligence—that is, law enforcement intelligence, business intelligence, and private intelligence—bear the hallmarks of their parent. The foundational features can be outlined as:

1. Decision makers should not base decisions on information but rather on intelligence;
2. The production of intelligence must be timely;
3. Intelligence strives to be accurate to win the confidence of those making decisions;
4. Intelligence must be usable; and
5. Intelligence products must be able to provide sufficient insight to enable decisions to be made.

Like military and law enforcement intelligence, national security intelligence is seen as the key to decision making. As Ransom put it: "Nothing is more crucial in the making of national decisions than the relationship between intelligence and policy, or, in a broader sense, between intelligence and action."[16]

National security intelligence is a function that is carried out by a country's foreign diplomatic service or sometimes its closely affiliated services. It also includes specialized agencies such as America's Central Intelligence Agency (CIA), Australia's Office of National Assessments (ONA), Canada's Canadian Security Intelligence Service (CSIS), New Zealand's Security Intelligence Service (SIS), and Britain's Security Service (MI5). These organizations centrally coordinate analysis through supporting arrangements with other agencies that process information collected from all sources—open, official, and covert.

National security intelligence could be considered a descendent of its military parent; although, the link between military and national security intelligence is at times so intimate that a clear demarcation cannot be realistically declared. It could be argued, therefore, that national security intelligence is not a descendent at all but a discipline that developed at the same time and in sympathy with military intelligence.

The responsibility of national security intelligence is to advise political leaders on the formulation of policies relating to a wide range of foreign policy and international political issues, including the discharging of responsibilities under a number of international pacts and treaties. To be sound and constructive, foreign policy (and where this realm overlaps with military strategy) must be based upon fact and realism. Many of the facts needed to support a nation's foreign policy are therefore provided by these intelligence agencies.

KEY WORDS AND PHRASES

The key words and phrases associated with this chapter are:

- Adversary;
- Agent;
- Applied intelligence research;
- Business intelligence;
- Counterespionage;
- Counterintelligence;
- Cover;
- Covert action;
- Decision maker;
- Espionage;
- Essential elements of intelligence;
- Information;
- Intelligence;
- Intelligence cycle;
- Intelligence requirement;
- Military intelligence;
- National security intelligence;
- Officer;
- Operational intelligence;
- Operative;
- Private intelligence;
- Strategic intelligence;
- Tactical intelligence; and
- Target.

STUDY QUESTIONS

1. By way of example, define the term *intelligence*.
2. Provide an overview of the five major types of intelligence as well as the different functions within the typology.
3. Describe what is different between the three major classifications of intelligence—strategic, tactical, and operational.
4. Identify two intelligence consumers in a military setting, and describe how they might use intelligence products.
5. What type of intelligence would it be if an analyst was tasked to assist field operatives to locate the individuals responsible for a terrorist attack on an iconic landmark? Explain why.

LEARNING ACTIVITY

Using Prunckun's article on comparative research,[17] investigate the reasons why the intelligence cycle is related to research cycles in other scholarly fields, like applied social research.

NOTES

1. Ira S. Cohen, *Realpolitik: Theory and Practice* (Encino, California: Dickenson Publishing, 1975), 41–42.
2. Terry L. Schroeder, *Intelligence Specialist 3 & 2*, vol. 1 (Washington DC: Naval Education and Training Program Development Center, 1983), 2-1.
3. Christopher Andrew, Richard Aldrich, and Wesley Wark, *Secret Intelligence: A Reader* (London: Routledge, 2009), 1.
4. Henry Prunckun, "The Intelligence Analyst as Social Scientist: A Comparison of Research Methods," *Police Studies* 19, no. 3 (1996): 70–72.
5. Ted Gup, *The Book of Honor: Covert Lives and Classified Deaths at the CIA* (New York: Doubleday, 2000).
6. T. J. Waters, *Class 11: Inside the CIA's First Post-9/11 Spy Class* (New York: Dutton, 2006), 4.
7. Ellis M. Zacharias, *Secret Missions: The Story of an Intelligence Officer* (New York: G. P. Putnam's Sons, 1946).
8. Harry Rositzke, *CIA's Secret Operations: Espionage, Counterespionage, and Covert Action* (New York: Reader's Digest Press, 1977), 119.
9. Dennis Fiery, *Out of Business: Force a Company, Business or Store to Close Its Doors . . . For Good!* (Port Townsend, Washington: Loompanics Unlimited, 1999).
10. Emergency Management Australia, *Operations Centre Management*, 2nd ed. (Canberra, Australia: Emergency Management Australia, 2001), 5–10.
11. Emergency Management Australia, *Land Search Operations*, 2nd ed. (Canberra, Australia: Emergency Management Australia, 1997).

12. Joseph A. McChristian, *The Role of Military Intelligence: 1965–1967* (Washington DC: GPO, 1974), 3.

13. McChristian, *The Role of Military Intelligence*, 3.

14. William Sammon, Mark Kurland, and Robert Spitalnic, *Business Competitor Intelligence: Methods for Collecting, Organizing, and Using Information* (New York: John Wiley and Sons, 1984), v.

15. David Kahn, *The Codebreakers: The Story of Secret Writing* (Toronto, Canada: The Macmillan Company, 1969), 360.

16. Harry Howe Ransom, *The Intelligence Establishment* (Cambridge, MA: Harvard University Press, 1971), 3.

17. Prunckun, "The Intelligence Analyst as Social Scientist," 67–80.

2

֍

The Intelligence
Research Process

This topic considers the process of intelligence research by examining:

1. Problem formulation;
2. Literature review;
3. Methodology;
4. Intelligence collection plan;
5. Data collection;
6. Data evaluation;
7. Quality control;
8. Purging files;
9. Data collation;
10. Intelligence systems;
11. Data analysis;
12. Inference development and drawing conclusions;
13. Making recommendations; and
14. Report dissemination.

Intelligence has to "educate its customers [but] this is a formidable task. . . . They have to be convinced of what it can and what it cannot achieve" by asking the right questions, at the right time and without flooding the system.[1]

PROBLEM FORMULATION

Problem formulation is the center of intelligence research. Aristotle is attributed as saying: "Well begun is half done," and this proverb resonates in the intelligence context.

But how do decision makers formulate their questions, and how do analysts arrive at the hypotheses that form the basis of their research projects? Because intelligence research is applied, the questions under investigation will have real-world origins. Whether the origin is geopolitical (national security intelligence), financial markets (business intelligence), criminal activity (law enforcement intelligence), or issues involving an adversary's order of battle (military intelligence), the questions decision makers pose are concerned with how to address the problem. Intelligence provides insight to guide possible options based on defensible conclusions derived from evidence-centered research.

> In the social and behavioral sciences, the term *hypothesis* is used, but sometimes in intelligence research, the term *explanation* is applied. In this book, the term *hypothesis* will be used.

Having said that, there are times when analysts are asked to provide decision makers with possible scenarios of what the future holds for a particular environment. In such cases, analysts are free to establish their own theories and create their own research questions. To do this, analysts use techniques like morphological analysis, which can generate large numbers of possible futures. Chapter 10 discusses how to perform morphological analysis.

While analysts can provide intelligence, the competing demands of tactical field commanders (e.g., executive directors in a business setting) and strategic decision makers can flood the intelligence system, thereby rendering it ineffective. Therefore, what is asked of an intelligence unit should be minimal and specific.

For instance, a single but nonspecific intelligence request can be quite counterproductive. Take the example of a field commander who asks to "see all the aerial photos relating to the terrain north of the Orrenabad Desert." This may result in boxes of assorted classified imagery being delivered hours or days after the time it is needed, rendering the information useless to the commander if he or she had intended to use it to plan an attack. Therefore, as specific requests assist intelligence analysts, they can reciprocate by yielding answers more useful to the decision makers when choosing the most appropriate operational options. Requests need to be specified using questions such as:

- How many troops does the enemy have positioned north of the Or-renabad Desert?;
- What is the enemy's order of battle?;
- Will the weather be favorable to launch a frontal assault on these positions over the next 24 hours?;
- If an attack is launched, can resupply of friendly forces be assured?; and
- Can air support provide both suppressing fire in the advance and evacuation of the wounded?

LITERATURE REVIEW

The purpose of the literature review is to seek research related to the issue under investigation in order to establish a conceptual as well as a theoretical context for the research project. The literature review places the analyst's project in the context of the wider issue. "No man is an island onto himself," wrote the poet John Donne, and his message, though intended for a different audience, applies to intelligence problems.[2] No research question exists in isolation to other issues. The literature review allows analysts to develop and ground their arguments, or simply to "tell their stories."

For analysts undertaking a new research project, a literature review is one of the first steps. Analysts need to discuss the theoretical base they intend to use to test their hypothesis. For law enforcement analysts, this might include deterrence theory, target hardening, rationale choice theory, differential association, social disorganization, or any number of other criminological theories.

For national security intelligence analysts, they might consider the use of sociological (e.g., Edwin Sutherland's theory on white collar crime[3]), anthropological, psychological, political science, economic (especially as they relate to illicit drug importation), business (in relation to organized crime and antiterrorism), or even military theories (e.g., counterterrorism and counterinsurgency) and apply them to the issue under investigation. The theory component of the literature review should have its own subheading (or some such literary device) so the reader does not have to "hunt" for this information.

The literature review should talk about why the issue being studied is a problem and how the theory (i.e., the presumed relationship between the variables), if applied to the problem, could help, improve, solve, make more efficient, and so forth. Effectively, it states what benefit the study's findings might have for decision makers or how the research results might "push back the frontiers of knowledge."

Analysts need to demonstrate they have an understanding of what research has already been conducted and how the new research will either add to it, address an area that has not yet been explored, or, if a different theory is being used, why changes might be expected that are yet to be realized. The key concepts and terms need to be discussed so they can be "operationalized" in the methodology.

It is unlikely the analysts will find research that is exactly the same as the inquiries they are making (unless they are conducting a reexamination of previously conducted research). From this perspective, there will be no issue so unique the analyst cannot locate some piece of related research to inform the current scholarly investigation.

As with other forms of scientific research, intelligence research seeks to push back the boundaries of knowledge by adding to what is known. Therefore, the literature review allows the analyst to introduce the problem under investigation expressed as a research question or hypothesis so the reader of the report can understand how it fits within the wider context, how exploration of the problem will provide understanding, and how to provide critical insight into solving it.

On a practical point, reviewing the literature allows the analyst to gain an appreciation of some of the problems encountered in previous research and, therefore, avoid repeating the same mistakes. These problems are often in methodological issues including inappropriate sample size or selection method, invalid measurement instruments, data reliability issues, or inappropriate statistical analysis (e.g., autocorrelation of time-series data is a commonly experienced problem).

The ancillary benefit of a literature review is the discovery of secondary data sources to be used in the study. For instance, an analyst may come across quotations from field notes or interviews, and these can be used as secondary sources of data.

Even though the literature review sets the scene and appears in the forefront of the research report, the analyst can always refer back to the literature in the results section of the report as a means of validating certain aspects of the findings.

METHODOLOGY

Arguably, this section is the most important aspect of the research process. If the analyst has crafted the research question well and placed it in its historical and theoretical framework, it will guide the rest of the research process. As the methodology deals with the tangible aspects of the study, the analyst will need to define the concepts being studied so they can be measured (i.e., operationalized).

The most popular research designs include evaluations (to plan intervention programs/operations), case studies (what is going on), longitudinal studies (has there been any change over time), comparisons (are A and B different), cross-sectional studies (are A and B different at this point in time), longitudinal comparisons (are A and B different over time), and experiments or quasi-experimental studies (what effect does A have on B).

The methodology requires analysts to identify the type of data they need to collect (whether these data are from primary and secondary sources or in the form of qualitative, quantitative, or both) and how these data will be collated and analyzed (e.g., statistically or content analysis) to test their hypothesis.

Analysts also need to consider related issues including sample size and control for confounding variables (i.e., the potential that what they are observing is due to something else that they are not measuring). These extraneous influences may pose limitations for their research (e.g., possible alternative explanations for the relationship between A and B) or limits inherent in the data.

INTELLIGENCE COLLECTION PLAN

Analysts use the intelligence collection plan as a conceptual tool to design and manage their data acquisition. The plan is a simple device structuring the analyst's thinking to develop a picture of the data and what is required, where it exists, and how it can be gathered. It outlines timeframes for collection of each piece of information against the timeframe of the research project, allows for noting when data have been collected, and allows for any items outstanding. Therefore, the intelligence collection plan can either be a simple collection tool or a combined collection and management tool. These can take the form of fishbone diagrams and data collection tables.

Because the objectives of individual intelligence agencies are so diverse, there are no set formats for collection plans. Nonetheless, a good collection plan should be couched in precise terms reflecting:

- The decision makers' *intelligence requirements* (IR) or *intelligence collection requirements* (ICR);
- The resources needed to collect data;
- What priority each data item has in relation to other data and within the whole data collection scheme;

- Who is responsible for collecting each data item; and
- The progress tracking of each data item.

The plan should be flexible to allow adjustment as changes in intelligence requirements emerge or if the objective of the research project alters.

In some agencies, there are two additional methods of acquiring information—the *statement of intelligence interest* and *essential elements of information* (EEI was discussed in the previous chapter). The former is a standing request to receive finished intelligence publications when released for dissemination. Usually, analysts register their interests in topics under investigation, and the reports are forwarded to them for consideration when they become available. The standing request can be viewed as an order that remains in force until cancelled, and as new material is lodged, it is sent to the analyst. EEI are time urgent requests for information (e.g., ground forces engaged in combat operations require EEI).

Fishbone Diagram

A fishbone diagram (see fig. 2.1) can be used to coordinate an intelligence collection plan. A fishbone analysis is usually conducted to identify and explore cause-and-effect issues but can be adapted by analysts to help manage the collection process.

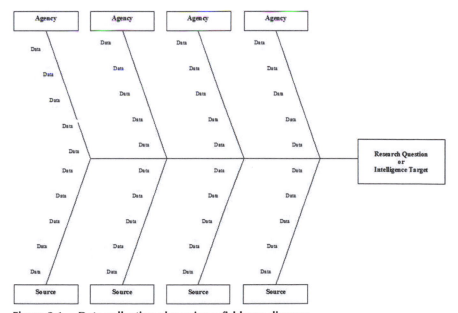

Figure 2.1.   Data collection plan using a fishbone diagram.

The intelligence target or research question is posted at the right-hand side of the diagram (the fish's head). The major bones of the fish are constructed by listing the different agencies or sources of information, and the minor bones subtend from the major bones, listing the data items required. As each piece of data is received, it can be crossed off the diagram with the effect of producing a visual aid as to overall progress. The pictorial information can be converted into a progress report in narrative form or as a statistical summary.

Data Collection Table

A data collection table can also be used to organize an analyst's data requirements (see table 2.1). Across the top of the table, the analyst lists the key issues for consideration. There are five shown, but other issues could be included in this row, depending upon the target and the ramifications associated with the research project. Other issues could include legal constraints, administration, communication, timing, and data security.

Starting with "Type of data," the analyst states the type of questions they need answered, placing each category in the row below. The analyst could substitute *type of data* with specific questions that need to be answered. This is followed by completing each cell by moving from left to right across the rows. Table 2.1 is an example for the aerospace industry compiled by a notional private intelligence firm.

Table 2.1.   Data Collection Plan in Table Form

Type of Data	Source	Risk	Expense	Priority
Statistical	Aerospace Facts and Figures	Nil	Cost of purchasing annual volume or online subscription	High
	Handbook of Airline Statistics	Nil	Cost of purchase	Medium
Manufacturers	World Aviation Directory	Nil	Cost of purchase	High
	Interavia ABC: World Directory of Aviation and Astronautics	Nil	Cost of purchase	Medium
Associations	Aerospace Industries Association of America	Nil	Cost of purchase	Low

DATA COLLECTION

Information can be gathered from a variety of sources. The diversity of these sources is exemplified by category in the following list. Many of the information sources are the same for the five different types of intelligence practitioners—national security, military, law enforcement, business, and private intelligence.

It is conceivable that a single piece of information could have application to each functional intelligence group (e.g., information relating to a terrorist cell could be of interest to law enforcement intelligence that have a responsibility to prevent and deter such attacks on the homeland). Likewise, the same information could be of interest to:

- National security and military intelligence if the cell is internationally based or operates from overseas;
- Business intelligence if the target of the terrorists is their industry or their facilities; or
- Groups employing private intelligence, such as the antinuclear lobby to highlight the vulnerable nature of a nuclear facility in relation to terrorism.

Law Enforcement Intelligence

- The public;
- Crime investigators (detectives);
- Patrol officers;
- Police records;
- The media;
- Businesses;
- Open source information (particularly the Internet);
- Government departments and agencies;
- Informants;
- Covert surveillance (physical and electronic);
- Undercover operatives; and
- Other law enforcement agencies and government departments.

Business Intelligence

- Open source information (particularly the Internet);
- A business's own internal records;
- Information supplied by other businesses;
- The media, trade, and other open source publications;

- Sales personnel;
- Customers;
- Distributors;
- Raw material and component suppliers;
- Government departments and agencies;
- A business's research and development section(s);
- University and independent research bodies;
- Market research surveys;
- Reverse engineering; and
- Covert physical surveillance (e.g., a hired private investigator).

Military Intelligence

- Open source information (particularly the Internet);
- Surveillance planes;
- Surveillance satellites;
- Electronic intercepts;
- Reconnaissance teams;
- Field operatives;
- Diplomatic missions and embassies;
- Defectors;
- Prisoners;
- University and independent research bodies; and
- Other government departments.

National Security Intelligence

- Open source information (particularly the Internet);
- Clandestine operatives (official cover);
- Covert operatives (nonofficial cover);
- Recruited agents;
- Diplomatic missions and embassies;
- Surveillance planes;
- Surveillance satellites;
- Electronic intercepts;
- Defectors;
- University and independent research bodies; and
- Other government departments.

Private Intelligence

- An organization's own internal records;
- Open source information (particularly the Internet);
- Information supplied by other organizations;

- The media, trade, and other open source publications;
- Staff;
- The public;
- Government departments and agencies;
- An organization's research section;
- University and independent research bodies;
- Surveys; and
- Covert physical surveillance (e.g., a hired private investigator).

DATA EVALUATION

Evaluating information is an integral step in the analytic process and usually takes place as information is gathered. The data are evaluated according to the reliability of the source and the validity of the actual information. When evaluating information, the analyst asks many questions including:

- How reliable is the information source?;
- Has the source provided information before?;
- How accurate is the information?; and
- How recent is the information?

With some types of intelligence, particularly national security intelligence and military intelligence, deception is a particular concern. In such cases, analysts need to evaluate data to distinguish between objective information and that tainted by bias.

> Secondary sources such as government press offices, commercial news organizations, [nongovernment organization] spokespersons, and other information providers can intentionally or unintentionally add, delete, modify, or otherwise filter the information they make available to the general public. These sources may also convey one message in English for US or international consumption and a different non-English message for local or regional consumption. It is important to know the background of open sources and the purpose of the public information in order to distinguish objective, factual information from information that lacks merit, contains bias, or is part of an effort to deceive the reader.[4]

The evaluation process firstly assesses the source's reliability and, secondly, the information's accuracy. In theory, this process is performed on each piece of information collected. However, in agencies collecting large volumes of data, this may be an automated process where a generic rating is assigned if the data are merely stored, but if used in an intelligence research project, it is reevaluated on an individual basis. Each piece of data is assigned an alphanumeric rating indicating the degree of confidence the analyst has in that piece of information. This system is universally known as the *admiralty ratings* and is shown in tables 2.2 and 2.3.

Table 2.2.   Information Accuracy Codes

	Admiralty Ratings	
Code	*Descriptors*	*Estimated Probability of Truth*
1	Confirmed	100%
2	Probably true	80%
3	Possibly true	60%
4	Doubtful	40%
5	Improbable	20%
6	Misinformation	
7	Deception	
8	Cannot be judged	50%

Imagine a field operative obtains a piece of information from an agent in place, but this agent is a new source that has never been exploited before. The reliability for this piece of information would therefore have to be F—the reliability cannot be judged. If the information obtained came from a database that has been the source of previous information (i.e., from another agent) and has proven to be truthful in almost every instance, then an accuracy code of 2 would be assigned. The combined code would be printed on the document to show its overall rating. Customarily, accuracy rating precedes the reliability code—for instance, F-2. Although the ratings must be logical, assigning a rating of E-2 (unreliable source but probably true) would raise questions about whether the evaluation process was rational.

Regarding the accuracy codes 6 and 7, the analyst should be cognizant they may obtain data that are unintentionally false, illogical, or contradicted by other sources. In these cases, a code of 6—misinformation—should be assigned. As for disinformation (code 7), this is data that are shown by other sources as deliberately false.

Although the admiralty ratings represent an objective position, it is derived through a subjective process as judgment plays a key role. When as-

Table 2.3.   Information Reliability Codes

	Admiralty Ratings	
Code	*Descriptors*	*Estimated Probability of Truth*
A	Completely reliable	100%
B	Usually reliable	80%
C	Fairly reliable	60%
D	Not usually reliable	40%
E	Unreliable	20%
F	Cannot be judged	50%

signing a rating, the analyst must consider such things as the accuracy of previous information provided by the source and the source's field capabilities (i.e., does the source have access and the ability to obtain what has been delivered?). Evaluation is a difficult process but an important one as personal or agency bias can adversely affect the results of an intelligence research project. A good evaluation is the result of the source's reliability being evaluated independently of the credibility of the information. Information discovered to be irrelevant to the issue under investigation should be disposed of according to the analyst's agency document destruction policy (e.g., shredding).[5, 6, 7]

QUALITY CONTROL

To assist the analyst in determining the relevance of any particular piece of information to the research project, a method of labeling each data item is needed. If the analyst can correlate data with other sources of information thought to be reliable, then, theoretically, this will increase the chances the data are correct. Some intelligence agencies require that to assign a data item an A-1 rating, the data need to be verified independently against two or more sources.

However sound this procedure appears, it can be subject to phantom or ghost data—data that are self-validating. By way of example, the author was once involved in an intelligence operation, and the operational team obtained document A, which cited the occurrence of a particular meeting involving one of the operation's targets. If true, this information would have led to a major breakthrough by piecing together a hitherto incomplete picture of a highly illegal industry.

So the intelligence team set about checking this information with other sources, and as a result, they came up with two independent confirmations, documents B and C. But acute observation told the team something seemed odd. There were a few words in documents B and C (the additional confirmatory documents) that were similar. Checking the sources (a time consuming task) revealed, to the team's astonishment, all three reports (A, B, and C) were based on a single report, which in turn was authored by another source! It should be noted that although these were documentary reports, the same holds for oral reports by an informant or undercover agent.

In this case, the data were self-validating. If an intelligence assessment was prepared on these data, the consequences do not need to be spelled out. But given the guidelines, the assessment would have been correct. It is worth noting that data validation systems, such as the admiralty ratings, should be only guidelines, not rules.

PURGING FILES

Another aspect of quality control is purging redundant data. Intelligence systems tend to accumulate data very quickly, and in turn, these data contain material which is not always central to intelligence projects. This is mainly due to the analysts' inability to predict what data will be needed in the future; therefore, if a piece of data could be used, it tends to be retained.

However, at some stage, an assessment of the agency's holdings should be done (i.e., at the end of a project or operation). During such an audit, material lacking accuracy, relevance, timeliness, or completeness should be purged from the system. Retaining this material does nothing but detract from the quality of the overall database, and if it is used in an intelligence assessment, the data will only reduce the accuracy of, and possibly harm, the final recommendations. In some jurisdictions, having irrelevant data could be a breach of the criminal law.

When data lacking these qualities are discovered, they should be subject to upgrading. But, if the cost of validating or verifying this information proves to be in excess of the potential value of the anticipated result, the data should be removed. This process should be done on a project-by-project or operation-by-operation basis. In this way, the validating and verifying is done with the most ease and effectiveness, while the project is still fresh in the minds of the analysts. Likewise, the accuracy of the index should be maintained daily—corrections and updates should be done as they are encountered. This makes for efficient and effective management of time and of the database.

A biyearly quality control exercise could be done on the whole database. The criteria cited in table 2.4 can be applied to the data and up-

Table 2.4. Some Suggestions for Retaining or Purging Information

Accuracy	How reliable is the source?	
Relevancy	Does it supply the intelligence unit with information necessary to complete its project?	
Timeliness	Does each piece of the information relate to the current project or operation?	
Completeness	Externally obtained data	Is the document's source stated clearly? If not, can it be determined and inserted to be made complete?
	Internally generated data	Is the source of the facts referred to in the document footnoted or otherwise differentiated? If not, can the reader be provided with the source or authority for these data items?
	Considerations for all data	Are inferences and comments made by analysts clearly indicated in the document so as not to be confused with verified facts?

graded and/or discarded. In some intelligence agencies, if data cannot be judged, the information is sent to the originator of the report who is asked to upgrade it or otherwise make comment on it. The upgraded data are then entered into the system, replacing the old. If it cannot be upgraded, the data are destroyed.

DATA COLLATION

Collating takes place after the data have been evaluated, whereby the analyst brings together the disparate pieces of information so the data can be subject to some form of analysis. The collation process also acts to remove irrelevant, incorrect, or worthless information, which may be collected due to error, misdirection, or compulsiveness. Such data should be destroyed immediately after being identified; otherwise, it will not only cause congestion in the intelligence database but also may place the analyst and the agency in legal jeopardy if a judicial officer tasked with overseeing the agency's operations deems holding such data to be contrary to law.

The remaining data are then stored to be retrieved easily by the analyst. If the data are in an unstructured form, this can be done by:

- Registering the information;
- Indexing;
- Cross-referencing; and
- Keywording.

These data can then be filed so that an electronic database can be interrogated by the analyst using one or more analytic tools. Typically, unstructured data are stored on a mainframe or a PC server that is accessed by the analyst's workstation computer (which in turn is likely to be networked with other workstations and servers within the agency).

INTELLIGENCE SYSTEMS

Basic Data Storage and Retrieval Concepts

Information is stored for the purpose of retrieval and not stored for the purpose of warehousing documents. This may seem obvious, but in practice, this is not always the case. Information retrieval is the selective and systemic recall of warehoused information; hence, the information must be stored logically, or it may become lost "in the system."

Storing and retrieving data effectively does require technology. At the center of many intelligence systems is some form of index. The purpose of the index is to facilitate the retrieval process, but by its nature, an index imposes limits on how information can be retrieved. A superior indexing system allows for subjective, intuitive searching that is fundamental to intelligence research. The most common types of indexes are those listed below, and a good intelligence database should be capable of producing all.

Author Indexes. People, organizations, corporate authors, government departments and agencies, universities, research foundations, and so forth.

Alphabetical Subject Indexes. Headings, subheadings, cross-references, and qualifying terms.

Keyword in Context. A system specific to computerized systems. It operates by selecting keywords appearing in the body of the text (e.g., the keyword *blackmail* in a corruption report).

Hierarchical Indexes. Data items are arranged in a hierarchy, starting with topics of general scope and progressing to more specific topics.

Permuted Title Indexes. Systematically rotating the words in the title of a document/file. The success of the permuted index depends on the accuracy of the original author (or where no title is given, e.g., *ad hoc* papers and memos, the collector, writer, or researcher) creating the title to reflect the content of the document or file. This may be difficult in the cases where a document/file covers a number of topics.

Sound Indexing. *Soundex*[8] is an indexing method based on sound where names are encoded for retrieval.

Soundex-Based Systems

Soundex is a way of searching data where the exact spelling is unknown. This is particularly valuable as names can be spelled several different ways. Because a soundex-based search seeks the phonetic spelling of a word, a search will usually be more reliable if carried out for single words rather than combinations of words, such as titles and phrases (depending on the software package used).[9] The soundex method of indexing involves four steps:[10]

1. Retain the first letter of the name, and drop all occurrences of *a, e, h, i, o, u, w,* and *y* in other positions.
2. Assign the following numbers to the remaining letters after the first: 1 = *b, f, p, v*; 2 = *c, g, j, k, q, s, x, z*; 3 = *d, t*; 4 = *l*; 5 = *m, n*; 6 = *r*.
3. If two or more letters with the same code were adjacent in the original name (before step 1), omit all but the first.
4. Convert to the form "letter, digit, digit, digit" by adding trailing zeros (if there are more than three).

Internet-Based Systems

Internet-based search engines use a two-step process for indexing material posted to websites. The first step is "web crawling" or "spidering." Here a software agent methodically browses or crawls the Web and makes copies of web pages by following the hyperlinks on each page. These copies then form the basis for the next step, which is indexing.

However, because the algorithms used to construct Internet search engine indexes are commercial-in-confidence, they cannot be reviewed here. Nevertheless, it is widely understood these algorithms are based on interdisciplinary concepts borrowed for the likes of linguistics, cognitive psychology, mathematics, and informatics (and perhaps others).

As Internet-based search engines are commercial ventures and the companies running these services have to compete in a very competitive marketplace, the owners try to balance the thoroughness provided by a search request with retrieval time. In doing so, the search engine needs to give the researcher the most relevant web pages. It is in deciding what is relevant that each search engine differs in its construction of its search algorithm—hence, the interdisciplinary and multidisciplinary approach to indexing.

DATA ANALYSIS

Information is analyzed to draw conclusions about an activity, a person(s), a group(s), or an organization(s) at the center of inquiry in order to provide insight for the decision maker (i.e., *information* is analyzed to produce *intelligence*). The analytical process can be described in a number of steps that comprise:

- Examining the collected data;
- Sorting facts from opinions (i.e., evaluation of the information);
- Developing inferences (by means of statistical or other methods);
- Discussing the strengths and limitations of the various inferences (e.g., based on probabilities); and
- Drawing conclusions from these results and making recommendations in accordance with the decision maker's intelligence requirements.

Despite the complexity of data analysis, the process of analyzing data is generally straightforward. Intelligence research projects mostly perform analyses following a three-step process:

1. Preparing the data by cleaning errors and anomalies that may have crept in during the collection phase;
2. Organizing the data so it can be described statistically if quantitative and in other ways if qualitative; and
3. Testing the research hypothesis (or model) using either statistical tests or specialized analytic techniques (depending on the type of data).

The first step—preparing the data—starts with some form of logging that allows the analyst to check what has been collected against the information collection plan. Once assured all data items are present, the analyst checks the data for accuracy, which involves entering the data into a software program. Depending on the type of data being processed, it may mean "double entering" the data to ensure there are no mistakes in data entry or checking that the data item is within a specified range or in a certain format. Software programs designed for data analysis will usually have "error trapping" sub-routes that highlight such errors or allow the analyst to set parameters depending upon the data being manipulated.

The purpose of using a software package is to provide some form of organization to the data so descriptive analysis can be performed. Most analytic software has this function and will produce a range of descriptive statistics forming the basis for testing the research hypothesis. Descriptive statistics are not only part of both quantitative research and qualitative projects, but also they are important as they describe what is going on. Such descriptions include graphical representations (e.g., pie diagrams, bar charts, or line graphs) coupled with a narrative discussing the various quanta.

INFERENCE DEVELOPMENT
AND DRAWING CONCLUSIONS

The final phase of the intelligence research process is testing the research hypothesis. This can be done using statistics or another technique if the data are qualitative. The overall purpose is to draw conclusions allowing the analyst to extrapolate meaning from the data to some level beyond the immediate. Based on the sample data, an analyst may form some inference that could be applied to the general population from which the sample was derived. Alternatively, an analyst may use inferential statistics to form a judgment about the probability that the observations regarding two groups are the result of some variable acting on one of the groups, and the difference would not have occurred if chance was the only factor.

An *inference* is a statement (or proposition or judgment) drawn from data that have been subject to some form of analysis. In this way, the statement follows logically (either deductively or inductively) from the data.

Inferences may be based on as little as one or two premises, or they can feature many premises in a cascading fashion depending on the data and the original research question. A simple example of an inference based on a deductive process consisting of two premises is:

- People are criminals because they have been found guilty of breaking the law.
- Mack DaKnife has been found guilty of breaking the law.
- Therefore, Mack DaKnife is a criminal.

You will note in the deductive reasoning process, the analyst moves from specific data items to a general position. In inductive reasoning, the opposite takes place—the analyst starts with a generalized position and moves to the specific, as seen in this example:

- Country Q is similar to Country X.
- Country Q provides a safe haven for terrorists.
- Therefore, Country X provides a safe haven for terrorists.

In the first example, the two premises are both true; therefore, the inference is valid. Although the premises in the inductive reasoning example are not false, there are many other particulars (i.e., variables) that need to be considered before the analyst can draw the inference that has been presented about Country X's providing a safe haven for terrorists.

The striking feature of the two approaches is that inferences developed by using deductive reasoning do not suffer the same degree of uncertainty as those developed by inductive reasoning. Nevertheless, inductive data analysis is useful in certain situations, such as exploratory qualitative case studies.

A general (logical) truth of a matter cannot be established by examining only one, or a few, of the variables as there are potentially a very large number (some might say infinite) of variables that need to be weighed in an inductive argument.

Contrast this with a deductive argument where the premises and their conclusion are so integrally related that if the premises are true, then the conclusion must also be true. For inductive reasoning to be valid, it requires all the initial premises to be true, and as this is not possible, an inductive argument can only establish a degree of likelihood or probability.

Deductive reasoning produces either a valid argument or an invalid argument. Inductive reasoning can only produce a cogent, or sound (i.e., probable), argument.

Interpreting information is a cognitive process based on general knowledge, life experience, commonsense, and data collected in relation to the
issues under investigation. This process involves identifying new issues
and postulating the significance of these issues, which might include being viewed from the perspective of the target.[11] The "so what?" questions
can be answered in the interpretative process: what does this information
mean in relation to the issue under investigation? The answers provide
a useful starting point to formulate future courses of actions and make
recommendations.

> It is equally important to discover evidence of disagreement in the data as
> it is to find evidence that supports the research hypothesis. Differences add
> density to the conclusions drawn. Therefore, analysts should look for outli-
> ers and rival explanations in the data.

MAKING RECOMMENDATIONS

Recommendations stem from the conclusions, and it is the point where
the analyst presents the decision maker with a proposed course of action.
The action stems directly from the original research question, or it may
stem from an ancillary issue the research has discovered.

At first glance, the process of making recommendations appears
straightforward—an action is made in the form of a proposal: it is recommended that ABC intervene by conducting an XYZ-style operation. Of
course, such a recommendation would be fleshed out to include, perhaps,
time frames and other considerations. However, a recommendation such
as this would more often than not be found inadequate by a decision
maker because it offers no options or choices.

It is rare that decision makers will base their resolves purely on factual
evidence. As strange as this must sound, decision makers balance the
facts gleaned from an intelligence report with the political imperatives
surrounding the issue being studied. Often social, cultural, historical, economic, and other considerations enter into the decision maker's thoughts
before embarking on a course of action.[12]

The art of recommendation-making is, therefore, another exercise in
intelligence research. Intelligence analysts are not policy makers nor are
they political advisors, but they are, nevertheless, subject experts who can
offer a range of options that can potentially address the original research
question—from a "do nothing" option (yes, doing nothing is a choice) to
a "Rolls-Royce standard" response (whatever that might be in the context

of the issue). In between these options will usually be two, three, or more options that can be presented on a scale of increasing efficiency or effectiveness (i.e., a projection of the potential impact or consequence of the options). Each option, as it climbs the scale toward the Rolls Royce standard, will usually carry a heavier price in terms of financial, personnel, or material resources and perhaps time or other cost factors.

Although it is not always the role of the analysts to cost each of these factors, it may be incumbent upon them to at least brief the decision maker as to the relative expenditure required for each option. They then have to present an estimate or rank them in order of magnitude. This serves as a benchmark for the decision maker to assess the options within the package.

Take for instance an intelligence report recommending a surveillance team be tasked to gather additional information about a criminal gang involved in the "rebirthing" of stolen motor vehicles. Based on the research and using deductive reasoning, this conclusion may be considered to be on firm ground. However, the agency may only have one full-time surveillance team, and tasking it to gather information may place an existing surveillance operation into the importation of heroin at risk.

The police commander needs to understand the risks as well as the costs in order to weigh which operation will be supported. The outcome may never be known to the analyst as the drug operation may be a classified project and the analyst has no need-to-know. Likewise, the final decision should not be seen as a slight on the analyst's work should the recommendation not be acted upon. It is a balancing act on the part of decision makers to allocate resources under their command as best they can. Nonetheless, it is important analysts are aware that they may be called upon to provide options and cost estimates so the best decision can be made.

Another way to describe how making recommendations fits into the intelligence research process is through the five Is:

Intelligence. Information gathering and analysis in relation to the problem, leading to making recommendations for intervention (options).

Intervention. The presentation of a set of responses or policy options addressing the proximate and/or distal causes of the problem.

Implementation. The action needed to translate the recommended interventions into practice.

Involvement. The enlistment of key partner agency(s) and/or stakeholder(s) to implement the recommended option.

Impact. The evaluation of intervention's outputs and outcomes.

The Straw Man Technique

One method to help guide decision makers to the preferred option is to use the straw man technique. A proposal or solution is drafted (i.e., the straw man) and placed among the range of alternative options. The analyst then discusses this option by criticizing the shortcomings of the alternative recommendations. In this way, the straw man provides a springboard for exploring other possibilities. Presenting recommendations in this way allows the decision maker to understand the strengths/benefits of the preferred option when contrasted with others. It can also facilitate some idea generation by the decision maker who may then request modifications to the preferred option based on political considerations that were beyond the privy of the analyst. Regardless, the straw man approach increases the chances that the decision maker will be more comfortable with and, hence, "own" the final choice.

REPORT DISSEMINATION

Dissemination is the term commonly used for the last phase of the intelligence cycle where the product (e.g., a report or briefing) is delivered to the decision maker. Because intelligence reports vary in size from one-page briefings (e.g., tactical or operational reports) to book length studies (e.g., strategic assessments), it is hard to categorically state who the end user will be within an agency or even at what level of political leadership it may ultimately end up.

In the case of business intelligence, an intelligence report may be received by a board of directors (i.e., nonexecutive level), or it might be considered by the corporate executives who make day-to-day decisions about the business. Then again, it may only be read by a division-level manager who simply needs the insights to plan production schedules.

In the case of national security intelligence, strategic reports, unless in the form of a national estimate (essentially a synopsis of a much larger document), will not be read by a decision maker. In all likelihood, strategic reports will only be read in their entirety by one of the decision maker's staff or advisers. These personnel will extract what they consider the key findings (and these may not be the key findings of the study) and brief the decision maker themselves. A strategic intelligence study may even be destined for a subject specialist, who may be a fellow intelligence analyst within another agency.

KEY WORDS AND PHRASES

The key words and phrases associated with this chapter are:

- Admiralty rating;
- Data analysis;

- Data collation;
- Data collection table;
- Data gathering;
- Deductive logic;
- Dissemination;
- Fishbone diagram;
- Inductive logic;
- Inferences;
- Information accuracy;
- Information evaluation process;
- Intelligence collection plan;
- Intelligence collection requirements;
- Intelligence systems;
- Methodology;
- Operationalize;
- Source reliability; and
- Straw man.

STUDY QUESTIONS

1. Identify two reasons for conducting a literature review, and explain their importance to the overall intelligence research process.
2. What are some of the benefits of data collation? List three issues, and describe the positive impact of each.
3. Why formulate an information collection plan? Discuss what could go wrong if a plan is not incorporated into an intelligence research project.
4. Explain the purpose of using a software package for collating data.
5. Outline three methods for indexing data. Then choose one and describe its strengths and its limitations.
6. Outline some reasons why an analyst should review data held in the agency database or file system for purging.

LEARNING ACTIVITY

Review the difference between inductive and deductive reasoning. Construct two simple inductive arguments and two simple deductive arguments. Provide a short explanation as to why each is so.

NOTES

1. Walter Laqueur, "Spying and Democracy: The Future of Intelligence," *Current* (March/April 1986).

2. John Donne, *Meditation XVII* (1622), http://www.online-literature.com/donne /409/ (accessed November 11, 2008).

3. Edwin H. Sutherland, *White Collar Crime: The Uncut Version* (New Haven, CT: Yale University Press, 1983).

4. U.S. Department of the Army, *FMI 2-22.9: Open Source Intelligence* (Washington DC: Department of the Army, 2006), 2–10.

5. International Association of Chiefs of Police, *Law Enforcement Policy on the Use of Criminal Intelligence: A Manual for Police Executives* (Gaithersburg, MD: IACP, 1985).

6. Jack Morris, *The Criminal Intelligence File: A Handbook to Guide the Storage and Use of Confidential Law Enforcement Materials* (Loomis, CA: Palmer Press, 1992).

7. Henry Prunckun, *Information Security: A Practical Handbook on Business Counterintelligence* (Springfield, IL: Charles C Thomas, 1989).

8. Soundex was originally developed by Margaret K. Odell and Robert C. Russell, cf. U.S. Patents 1261167(1918) and 1435663(1922).

9. Henry Prunckun, *SpyBase* (Adelaide, Australia: Slezak Associates, 1991).

10. Donald E. Knuth, *Sorting and Searching*, vol. 3 of *The Art of Computer Programming* (Reading, MA: Addison-Wesley, 1973), 391–92.

11. Michael Scheuer [pseud.], *Through Our Enemies' Eyes: Osama bin Laden, Radical Islam, and the Future of America* (Washington DC: Brassey's, 2002).

12. Terry-Anne O'Neill, "The Relationship between Intelligence Analysis and Policymaking—Some Issues," *Journal of the Australian Institute of Professional Intelligence Officers* 8, no. 1 (1999): 5–22.

3

❧

Scientific Methods in Intelligence Research

This topic discusses how scientific methods are used in intelligence research by examining:

1. Scientific research methods;
2. Reasoning;
3. Probability;
4. Hypothesis testing;
5. Constructing a hypothesis;
6. Variables;
7. Operationalizing variables; and
8. Measuring variables.

SCIENTIFIC RESEARCH METHODS

Scientific methods are the techniques and processes used by researchers to:

- Investigate social, psychological, political, military, business, and economic phenomena;
- Acquire new knowledge in these areas; or
- Set right or incorporate previously gained knowledge using the same.

It involves the systemic analysis of phenomena and logical problem solving. The aim is to provide insight through a transparent process in order to reduce uncertainty.

Reducing uncertainty is a problem well noted by military commanders like the Prussian general Carl von Clausewitz (1780–1831), who articulated such concerns in his landmark treatise *On War*.[1]

The methods are considered *scientific* because they are based on empirical evidence that can be observed (directly or indirectly) and measured. These data are subjected to the established principles of logic and must be repeatable, as in the physical world (e.g., chemistry and physics). Scientific methods of inquiry are founded on the steps of the intelligence cycle—problem formulation, data collection, data collation, analysis (including hypothesis testing), and dissemination.

Although the focus of intelligence research differs from other fields of inquiry, the features of scientific inquiry remain the same. Like other researchers, intelligence analysts pose hypotheses to explain phenomena and plan methodological approaches to study real-world problems.

One prominent feature of scientific inquiry is the objectivity of the method, so bias is reduced when collecting and interpreting the data. In academic fields such as sociology, criminology, psychology, history, anthropology, political science, and economics, the findings are shared through publication in scholarly journals and professional conferences. There are a few reasons for sharing the research results, including:

- Knowledge is cumulative, and all scholar researchers benefit from publication (i.e., increasing the sum of knowledge); and
- Colleagues can critique the methods and interpretations of a study to exercise some level of quality control through peer review.

But this is not the case with intelligence research as it is secret research. The audience is key decision makers responsible for protecting and preserving a way of life for many people; hence, the research is not intended for public airing, even within academic circles. For instance, the inability to critique methods and interpretations of a study is a drawback with intelligence research, but in its place, analysts could consider finding ways within their agency to have their work peer reviewed. This requires more than proofreading by a supervisor but could be done through a work-in-progress forum of fellow analysts where the researchers present their methods and findings.

REASONING

Deductive Logic

Deductive reasoning (or deductive logic) uses arguments to move from general statements (the *premise*) to a specific position to draw a conclusion. The key feature of deductive reasoning is that the premises used to create the argument must be true. The premise consists of one or more *propositions* as well as another proposition referred to as the *conclusion*. Because the premise is true, the conclusion must also be true.

To assess whether a deductive argument has been constructed correctly, the analyst must ensure the argument is *valid* and *sound*. An argument is valid if the conclusion is a logical consequence of the premise. It will be sound if it is valid as well as having a true premise. Consider the following example:

- All people breathe air (the major premise).
- The president of the United States is a person (a minor premise—there may be others).
- Therefore, the president breathes air (the conclusion).

Note, however the *validity* of a deductive conclusion is not affected by the fact that the premise is not true as the conclusion will still be valid. Consider this example:

- All drug addicted people live on boats.
- All people are drug addicted.
- Therefore, all people live on boats.

Although this argument is *valid*, it is not *sound*, and for the argument to be sound, the conclusion must be based upon a premise that is true. But analysts will not always have both validity and soundness in their arguments, and if they did, there would be little cause to conduct research. When contrasting deductive logic with inductive logic, one finds rather than the arguments moving from general statements to a specific position to draw a conclusion, the reasoning moves from the particular to the general, and this is usually the basis of intelligence research.

Inductive Logic

Inductive reasoning makes generalizations based on individual observations (e.g., phenomenal patterns). It is used to attribute traits or relationships, or to formulate rules or theories. For example:

- Premise: There is a shortage of heroin for sale on the streets of Sydney.
- This infers the general proposition or major premise: There is a shortage of heroin throughout all of Australia.

Although inductive reasoning is valuable for posing a hypothetical position, it cannot be seen in the same light as deductive logic—both valid and sound. For the conclusion to be sound, the premise needs to be true, and it is not likely an analyst will have every piece of information to construct such an argument, so the premise is true as the number of particulars may be in the hundreds or thousands.

> *Belief* is like *faith*; both are expressions of opinion or conviction that something is true but have not been subjected to rigorous proof based on evidence. Belief and faith are, therefore, not recognized research methodologies.

PROBABILITY

Because inductive logic can be uncertain, analysts use probability when drawing conclusions: "this is the most likely truth based on what is presently best evidence." An analyst who tests a particular theory will construct a hypothesis based on the inductively produced theory; then using an appropriate method, the analyst will subject the research question to scientific inquiry.

Consider a study into a heroin shortage in Australia where illicit drug users in a few areas of Sydney experienced difficulty in obtaining heroin. A number of direct and indirect measures of availability were collected, and these data were subjected to analysis. The conclusion was that there was less heroin on the streets.

Some scholars extrapolated from these particulars an argument that the shortage was due to law enforcement efforts.[2] The conclusion was valid, but the argument failed to consider a number of other propositions making the premise not true, including a Taliban-imposed restriction in Afghanistan where opium is grown and diversion of Southeast Asian heroin (destined for the Australian black market) to Europe.

The initial conclusions were no doubt valid, but their soundness was questionable. It is likely that law enforcement played some part in the heroin shortage, but the probability that it was the sole basis for such results was highly unlikely. Therefore, what analysts need is some means of assigning a level of probability to the conclusions, including statistical techniques.

Tests of statistical significance can determine the likelihood that a particular result could have occurred by chance. Chance occurs when forces other than the independent variable act on the dependent variable. If it is unlikely that the result occurred by random variation, then the analyst can conclude the independent variable was responsible. The acceptable levels of probability (significance level) are 5 in 100 ($p < 0.05$) and 1 in 100 ($p < 0.01$) (this refers to the probability of rejecting the null hypothesis—see "Constructing a Research Hypothesis"). Therefore, if these results are obtained, the analyst can conclude that the study was significant at the acceptable levels, thereby supporting the hypothesis.

While numeric (quantitative) data can easily be used to conduct statistical tests of probability, analysts can also test nominal and categorical (qualitative) data using the chi-square. This technique will be discussed in chapter 8.

HYPOTHESIS TESTING

A hypothesis is the statement putting forward a proposition about a fact or set of facts. This statement, constructed by the intelligence analyst, is the basis of the research question or statement of guiding principle. Although a hypothesis can be created several ways, it is most likely the result of some form of inductive reasoning.

A problem is identified and a hypothesis framed, which usually starts out as the research question. The question might ask:

- How much can we expect organized crime to grow in three years' time?;
- What industries will organized crime be expected to engulf in the coming five years?; and
- Will their methods of operation differ from those being used today?[3]

These questions, as they stand, are too broad to test using the scientific method. They are tentative questions derived from the issues raised by decision makers, social observers and commentators, or the subject literature. These questions may form part of the rationale for the study, but they need to be refined to be testable and falsifiable. Take for instance the question that asks: what industries will organized crime be expected to engulf in the coming five years? The analysts might postulate that the nightclub industry may be the target of organized crime based on observations of local or interstate field operatives or the experiences discussed in the literature from overseas. Using this theory, they craft a testable hypothesis to be the subject of their inquiries. It can be in the form of a question, a statement, or an "if" statement, for example:

- Will the nightclub industry be the target of organized crime?;
- Nightclubs are an attractive target for takeover by organized crime; or
- If organized crime is looking to expand its influence, then the nightclub industry will be a target.

Any of these hypotheses are acceptable as they allow the analyst a clear focus for study. Each needs to consist of a dependent variable and an independent variable to allow the former to be manipulated by the latter. As a hypothesis must be testable, it must also be falsifiable. Consequently, a hypothesis can never be "proven," but it can be "supported," and it is to what degree a hypothesis is supported that determines its acceptance. So, statements such as: "and this is strong evidence in support of the hypothesis" or, alternatively, "these data do not support the hypothesis" (i.e., do not reject the null hypothesis) are common when discussing research findings.

CONSTRUCTING A RESEARCH HYPOTHESIS

To construct a research hypothesis, analysts must state what they consider their research will confirm.

Intelligence Hypothesis (H_1). Nightclubs are an attractive target for takeover by organized crime.

A study may have more than one hypothesis, so each hypothesis is labeled H_1, H_2, H_3 and so forth. Then the analyst must construct a null hypothesis (H_0), which is simply the reverse of the research hypothesis.

Intelligence Null Hypothesis (H_0). Nightclubs are not an attractive target for takeover by organized crime.

A null hypothesis is established to provide a form of devil's advocate for the analyst. There is much data that can falsely support the research hypothesis; therefore, the analyst must aim to reject the null hypothesis. If the null hypothesis cannot be rejected, the analyst can state there is support for the research hypothesis. This is a subtle but important distinction to understand.

One technique that is useful for refining a difficult problem is that of *restating the problem*. Consider the problem of traffic accidents for the police: how can police officers reduce motor vehicle accidents? Police may try to do this by increasing the use of radar and mobilizing high-visibility patrols.

Suppose then that the problem is restated as: How can police make motor vehicle travel safer? Then solutions that present themselves widen considerably. Additional options could include driver education programs, working with government departments that oversee roads and highways to identify dangerous intersections, turns, or "blind spots." It might also include issuing safety information to pedestrians, bicyclists, passengers, young drivers, aged drivers, truck drivers, and motorcyclists.

Restating the problem has the advantage of highlighting different goals, and these new goals present different solution paths. The more solution paths there are for a problem, the more likely it is that a solution can be found to treat it.

VARIABLES

A variable can take many categorical forms: items, actions, thoughts, mindsets, and moments in time or any number of other category types the analyst is trying to measure. To create a testable hypothesis, the analyst needs an *independent variable* and a *dependent variable*.

The independent variable is independent or not affected by other variables in the study. It acts on the dependent variable, which is the subject of the inquiry. For instance, nightclubs are the independent variable in the example given above about the problem of invasive organized crime.

The dependent variable is, therefore, the factor that depends on other factors—in the current example, organized crime. A simple way of remembering the two variables is to repeat this phrase: the [independent variable] causes a change in the [dependent variable]. To illustrate the point: [nightclubs] cause a change in [organized crime]. Any number of industries can be substituted for nightclubs to measure if there is any change in organized crime (i.e., referring back to the research question of being "an attractive target for takeover by organized crime").

In addition to the distinction between these two variables, it should be noted that all variables have *attributes* or specific values. The variable *gender* has two attributes: *male* and *female*.

Moreover, the attributes associated with each variable needs to be *exhaustive*, so the list should include all possibilities. For instance, if the variable of *religion* is used, then limiting the attributes to Catholic, Protestant, Jewish, and Muslim would be a mistake as there are many more religions being practiced. One way to address this issue is to expressly cite the major attributes but then use a category like "other" to account for the remainder.

Finally, attributes need to be *mutually exclusive*, so no variable should have two attributes simultaneously. Even though this may seem self-evident in practice, achieving this is not so simple. If a crime analyst is seeking to interview prisoners in jail to gauge their desire to join a "gang," then the variable "gang status" with only two attributes "member" and "nonmember" would be inadequate. How would the analyst record an answer where a prisoner is a member of a gang but is also seeking membership of a second gang (or wants to quit his present gang and join another)?

OPERATIONALIZING VARIABLES

Operationalizing variables refers to defining the variables in terms so they can be observed and measured. This is necessary because when a phenomenon can be observed, it can be measured, and this measurement can lead to management. But not all variables are easily operationalized. Consider how one would define:

- Organized crime; or
- A nightclub.

Variables that are objective, effort independent (i.e., involuntary), or concrete are easier to observe and, therefore, easier to measure. Those that are subjective, effort dependent, or abstract are much more difficult to operationalize. One way of reducing the confusion (and potential source of error) is to search the literature for other studies that have used the same or similar variable and consider the appropriateness of adopting that definition. Reliability (consistency in measuring) and validity (the degree to which the data collection instrument can measure what is intended) of the operational definition is paramount in the scientific method as the same results should be obtained if the study is repeated. The process of operationalizing a variable involves three steps:

1. Define the concept to be measured;
2. Assess the best quantitative measures for the concept (as there can be several measures for different concepts); and
3. Consider the most appropriate method for obtaining these data (e.g., covert, unobtrusive, or open-source).

MEASURING VARIABLES

There are three methods of measuring variables—directly, indirectly, and through constructs. An example of a directly measurable variable (or an *observable variable*) is the number of people attending a diplomatic function hosted by a target country's government.

An indirectly measurable variable (sometime termed a *hypothetical variable* or an *indeterminate variable*) would be the income of the guests at the diplomatic function. This could be used to assess the position they hold within society.

A variable "observed" through constructs cannot actually be physically observed as it is an *abstract theoretical variable*. They are created to represent some phenomenon like the emotional reaction of witnesses to a suicide bomber.

Units of Analysis

An important concept to consider when observing and recording data is the *unit of analysis*, which is an entity to be analyzed in the course of the intelligence study. A quick way to grasp this is to view an example from the military:

- soldier;
- squad (or section);
- platoon;
- company;
- battalion;
- regiment;
- brigade;
- division;
- corps; and
- army.

It is termed a *unit of analysis* because it is the level at which the study will analyze the data that determines the unit. For instance, if an intelligence study is comparing the type of weapons carried by insurgents who operate from two locations, the unit would be the individual insurgent. However, if the study was comparing how the insurgents engage friendly forces, then the unit of analysis might be at squad or platoon level. Despite this, it is not uncommon for studies to analyze data at several units of analysis.

KEY WORDS AND PHRASES

The key words and phrases associated with this chapter are:

- Attribute;
- Conclusion;
- Dependent variable;
- Hypothesis;
- Independent variable;
- Mutually exclusive;
- Null hypothesis;
- Operationalize;
- Premise;
- Probability;
- Proposition;
- Reasoning;
- Scientific;
- Scientific methods;
- Significance level;
- Sound;
- Source reliability;
- Unit of analysis;
- Valid; and
- Variables.

STUDY QUESTIONS

1. Explain why intelligence research employs the scientific method of inquiry.
2. What are the two forms of reasoning, and how would each be used in practice?
3. Outline the role probability plays in research.
4. What are the two integral parts of a properly formed hypothesis, and why is it constructed this way?
5. Describe the difference between the independent and dependent variables.

LEARNING ACTIVITY

Consider the issue of operationalizing variables; then define the variable "assaults" so that it can be observed and measured.

NOTES

1. Carl von Clausewitz, *On War*, trans. J. J. Graham (New York: Alfred A. Knopf, 1993).

2. Hank Prunckun, "A Rush to Judgment?: The Origin of the 2001 Australian 'Heroin Drought' and Its Implications for the Future of Drug Law Enforcement," *Global Crime* 7, no. 2 (May 2006): 247–55.

3. Henry Prunckun, *Special Access Required: A Practitioner's Guide to Law Enforcement Intelligence Literature* (Metuchen, NJ: Scarecrow Press, 1990), 3.

4

❧

Approaches to Intelligence Research

This topic provides an overview of the approaches to intelligence research by examining:

1. Quantitative research;
2. Qualitative research;
3. Mixed methods research; and
4. Intelligence research designs.

INTRODUCTION

Data are central to intelligence research, whether they are from primary sources, such as field observations, or from secondary sources, like numeric data in its various guises. This chapter examines the three paradigms for collecting, recording, and analyzing data: qualitative research, quantitative research, and mixed methods.

The difference between qualitative and quantitative research is akin to the human body where quantitative research would be the skeleton, and qualitative research is the flesh. One paradigm complements the other but does not necessarily replace the other. So, mixed methods combine both aspects, providing a fuller picture.

Each research paradigm has strengths and limitations. Sometimes the two approaches have been used jointly to *triangulate* research findings (i.e., increase the credibility of the study's results). However, in practice, the use of one over the other often comes down to the researcher's academic background and personal preference. Other times, it is simply a matter of how the research question is framed or what data are available.

QUANTITATIVE RESEARCH

Quantitative research rests on the analyst's making observations that can be measured using a measurement *instrument* (or *tool*). These instruments can be well developed and established as a psychometric test for vocational aptitude or a quickly designed survey. Some intelligence research projects lend themselves to a quantitative approach where the analyst:

- Is working with large data sets or collecting data from a large number of human subjects (especially if funds and support staff are few or timeframes tight);
- Has access to previously developed and tested data collection instruments that are applicable to the project;
- Has a group of decision makers who are more comfortable with numeric data or must demonstrate in measurable terms why a certain recommendation was made; or
- Is intending to estimate or predict some future possible outcome or event based on a sample.

QUALITATIVE RESEARCH

Qualitative intelligence research has the potential to extend understanding by suggesting tentative causal explanations. The analyst approaches the problem under investigation by collecting data in an unstructured way—there are no standardized questionnaires or limiting "boxes to check." Although some qualitative research may use a set of uniform questions to ask each respondent, the answers are not structured in the close-ended way that is a feature in quantitative research (e.g., characterized by simple "yes" or "no" answers or Likert-scale responses).

Qualitative research can be described as interactive field research or noninteractive documentary research. Data are collected by direct observation in the field or indirectly via diaries, journals, interviews, or focus groups. Secondary sources of data are numerous—documents exist in a myriad of forms. In intelligence research, qualitative data may also form the basis of a pilot study or develop a theory (e.g., grounded theory research) that could later be tested using quantitative data (e.g., using unobtrusive methods).

The keystone to qualitative analysis is *impression*. This is the effect the data have on the researcher who forms a view, image, or opinion about what the data "say." In a practical sense, the analyst examines the data and forms a judgment; then through a process called *coding*, they describe these impressions as concepts, categories, and properties (*dimensionalizing*). Following on, using one or more analytic techniques, which can include quantitative methods, the analyst then takes these categories and puts them back together in a way that makes connections; it is these connections that produce meaning. Some of the analytics used to make sense of unstructured data are described in detail in chapter 10—"Advanced Analytic Techniques."

An example of using both qualitative and quantitative techniques is in content analysis where the analyst will discuss themes that appear in the text as well as the reading ease and grade level required for adequate understanding. The analyst's impressions of the phenomenon under investigation are usually found in the section of the report that deals with conclusions.

Some intelligence research projects lend themselves to a qualitative approach when the analyst:

- Is interested in the target's behaviors, emotions, or thoughts;
- Has found little in the way of previously published information on the issue, and an overview is needed; or
- Needs an in-depth understanding of the issues (and perhaps associated issues) that quantitative data would not reveal (i.e., there needs to be more "flesh on the bones").

MIXED METHODS RESEARCH

Mixed methods research is a combination of the quantitative and qualitative paradigms. If an analyst is uncomfortable with one or the other, mixed methods is not only a good way of bridging the rift between the two, but also there are indications mixed methods research (sometimes termed *methodological pluralism* or *methodological eclecticism*) frequently results in superior research when compared to mono-method research.[1]

INTELLIGENCE RESEARCH DESIGNS

The most widely encountered research designs in intelligence research include the following:

Experimental

Experimental research is used to test for causal process where there is the ability to control the variables. In an experiment, there may be one or several independent variables that are manipulated to assess their effect on the dependent variable.

Quasi-Experimental

Quasi-experimental research are experimental research designs lacking all the characteristic of a true experiment—notably random selection. Quasi-experimental designs are often the form of time-series analysis, which can be either an interrupted time series or a non-interrupted time series (see Time Series Studies, later in this section). This approach lends itself to analyzing archival and other forms of unobtrusive data.[2]

Case Studies

Case studies are studies of single issues or problems and can be manifested in a person, a group, an incident, or an event. It is a systemic way of examining a problem extending beyond the use of a limited number of variables by providing an in-depth investigation into the target phenomena. Case studies can be single or multiple cases and need not be solely qualitative. Instead, they can use a quantitative paradigm or a mixed approach. This type of research design is well suited to strategic intelligence projects (see, for example, the case study into the 1986 Libyan air raid by the U.S. air force in retaliation of terrorist bombings in Europe targeting American interests[3]).

Evaluations

Evaluation research is the systematic assessment of an intelligence operation, a tactical service, or a strategic program. It is usually divided into assessments that are either formative or summative. That is, formative evaluations assess whether the operation, service, or program has been or can be improved through inputs, technology, training, or procedures. Summative evaluations assess whether the desired outcome of the operation, service, or program was achieved and if it was not (or partially), suggest improvements it might make to achieve its goals in the future.

Focus Groups

Focus groups are used in qualitative research to gather data from a large number of people simultaneously through open-ended questions;

although, closed questions can be asked to clarify points raised. One strength of this approach is the interactive discussion generated between the participants, creating information-rich results. This discussion is usually recorded by the researcher and transcribed later to derive data for the analysis.

In-Depth Interviews

In-depth interviews are similar to focus groups; although, there is only a single or possibly two respondents. The interview follows an unstructured format; however, the analyst will pose a basic set of open-ended questions to promote discussion and from which the data will be elicited to answer the research question. In-depth interviews are well suited to investigations where personal, sensitive, or confidential (e.g., classified) information is sought, making a group format inappropriate (e.g., counterintelligence investigations).

Ethnographies

Ethnographies seek to answer questions concerning the way people live. Ethnographic research is ideal for examining the relationships between culture and behavior (e.g., insurgencies) and is a good paradigm for gaining insight into an issue that could not be discovered using other research methodologies. While ethnography is a strategic, exploratory method, it can be used in a tactical or operational setting to answer specific questions.

Grounded Theory

Grounded theory is a research method where theory is developed from the data rather than research conducted to test an established theory. It is a truly inductive approach, taking the analyst from the specific to the general. In grounded theory research, concepts are the important elements of the analysis as these are used to develop theory (i.e., conceptualization of the data).

Time Series Studies

Time series studies are sometimes referred to as *repeated measures* studies because there are two or more observations (e.g., measures) taken from the same variable separated by time. Time series analysis aims to understand phenomenon represented by a sequence of observations and forecast future values of the variable.

Pre- and Post-Designs

Pre- and post-designs are usually used to measure changes caused by some form of intervention (i.e., a purposefully imposed independent variable on the dependent variable). Measurements are taken before the intervention (baseline data) and then after to assess the causality of the intervention. These types of studies are sometime termed A-B designs, where A represents the baseline phase and B the intervention. Variations to an A-B design are A-B-A, A-B-A-B, and B-A-B. Although it might seem odd to use a B-A-B design, it has its place in cases where the analyst comes across an issue that is in the process of having an intervention applied (B phase), and no baseline data were previously obtained. In such a situation, the analyst can request secession to the intervention in order to measure the post-impact (A) and then reintroduce it (B).

Meta-Analysis

Meta-analysis is a statistical research method that summarizes previously conducted studies. Analysts, rather than survey respondents, survey previously published research reports. It is, therefore, a purely quantitative approach, comparing the statistical results across a number of studies.

KEY WORDS AND PHRASES

The key words and phrases associated with this chapter are:

- Coding;
- Dimensionalizing;
- Impressions;
- Information rich;
- Instrument;
- Methodological eclecticism;
- Methodological pluralism;
- Mixed methods;
- Qualitative research;
- Quantitative research;
- Repeated measures;
- Research designs;
- Tool; and
- Triangulation.

STUDY QUESTIONS

1. Describe the difference between quantitative research and qualitative research.
2. Discuss the strengths and limitations of using quantitative research and qualitative research singularly.
3. Describe how these two approaches can be used to complement each other when used in combination.

LEARNING ACTIVITY

Consider the various research designs that can be used in intelligence research. Select one, and discuss how this design could be employed to research Country Q, which has recently made it public that it is developing a ballistic missile capability.

NOTES

1. R. Burke Johnson and Anthony J. Onwuegbuzie, "Mixed Methods Research: A Research Paradigm Whose Time Has Come," *Educational Researcher* 33, no. 7 (2004): 14–26.

2. Richards J. Heuer, ed., *Quantitative Approaches to Political Intelligence: The CIA Experience* (Boulder, CO: Westview Press, 1978).

3. Henry Prunckun and Philip Mohr, "Military Deterrence of International Terrorism: An Evaluation of Operation El Dorado Canyon," *Studies in Conflict and Terrorism* 20, no. 3 (July–September 1997): 267–80.

5

🌀

Unobtrusive Data Collection

This topic examines one of the most commonly used techniques for collecting data in intelligence research by looking at:

1. Benefits of unobtrusive methods;
2. Indirect data collection techniques;
3. Content analysis; and
4. Secondary analysis of data.

BENEFITS OF THE METHOD

Unlike surveys, in-depth interviews, and other methods where the researcher has direct communication with the researched, unobtrusive methods attempt to extract data without gaining the target's attention. Unobtrusive methods are conducted without the target knowing about an operation to collect information about him or her.

For intelligence analysts, there are a number of advantages for using unobtrusive methods in their research. Chief among these is what the analyst observes has actually taken place—as opposed to self-reporting by other techniques. These methods are, by nature, intrinsically safe as they do not place researchers or field operatives in contact with the target or in an environment that could be hostile or dangerous. They are, therefore, discreet and nondisruptive techniques.

Because these methods do not rely on direct contact, they increase the reliability of the study because the research can be replicated, thus allowing checks of the study's reliability and validity to be confirmed. Access does not present a problem because permission is not required, and in the case of a hostile target where permission could not be secured, the analyst does not have to resort to court-approved covert techniques.

Unobtrusive methods are cost efficient as undercover operatives do not have to be tasked, nor do surveillance teams or investigators/interrogators have to collect data. Because of its relative economy, unobtrusive methods lend themselves to being used for longitudinal studies where the analyst needs to follow the activities of a target over time.

Three techniques will be examined here—indirect data collection, content analysis, and secondary analysis of data. Arguably, these three comprise the most favored techniques by analysts as summed up by Professor Harry Howe Ransom, who wrote in his influential work on intelligence: "95 per cent of peacetime intelligence [comes] from open sources."[1] Professor Ransom's own analysis of the United States' national intelligence collection effort stated that in excess of 80 percent came from "overt, above-board methods [that] would normally be available to anyone with a well-organized information gathering system."[2]

This, by no means, should underestimate the value of covert and clandestine data collections and data from secret sources, but it highlights the relevance open source information and unobtrusive analytic techniques have in collection plans.

Analysts should never overlook the time-honored library. This is a rich source for open source information. Location-dependent, the quality and quantity of a library's holdings will vary, but generally, libraries have extensive holdings of nonfiction reference works as well as access to the interlibrary loan program. The latter provides access to nationwide holdings through member libraries.

INDIRECT DATA COLLECTION

Indirect data collection can be used as the analyst's primary source of data or used to supplement other forms of data in an attempt to triangulate results. Indirect methods lend themselves to collection by automated and electronic means.

The best way to understand indirect methods is by way of example. For instance, if a study wanted to measure the current interest in neo-Nazi issues (e.g., if there was some indication of a resurgence in hate crimes), an intelligence agency could set up a website to gather statistical information about those who log on, with the assumption being that those visiting such a website are either ideological supporters of this form of dogma or participants—potential or otherwise—of the ideology's edicts.

Data, such as the country where the Internet visitor originated, could be tracked along with the day, time, and web pages accessed. How long they stay on the website could be measured and whether they were returning viewers or once-only readers. Depending on the construction of the website, many other types of data could be collected, including provision for interactive input by the website visitor. If the agency desired, this data collection method could also be used as part of a sting operation where email addresses are obtained via an opt-in facility or a blog.[3]

The study of radio station listening preferences is an example relating to business intelligence that has been cited in the literature from time to time. In this study, the researchers are reported to have conducted an unobtrusive survey of radio stations that were favored by car drivers in a particular geographic area. The researchers attended automotive repair shops, and while customers' cars were being serviced, they noted the current radio station displayed on the radio (whether it was in the AM or FM broadcast band and what particular frequency).

There are other sources of indirect information, including monitoring radio and television broadcasts for speeches made by political leaders and field operatives attending public speaking forums. In the case of the latter, intelligence analysts need to brief and debrief operatives regarding the specific information required (as per the data collection plan), as well as any additional information gleaned by observation.

The advantage of this method is that it does not rely on the respondent having to tell the truth. A respondent who admits he or she listens to a station that plays music not in popular fashion or broadcasts certain political messages or is owned by a particular religious organization may feel embarrassed about his or her behavior. Prevention of embarrassment may lead respondents to lying, thereby distorting the data. Some of the limitations are that it only captures those respondents who can afford to have their cars serviced at a garage, thus excluding those who service their cars themselves (out of choice or necessity). It also assumes that the current station is the one most often or exclusively listened to.

Nevertheless, if a company wanted to market its goods or services to a demographic consisting of Mercedes-Benz owners, then attending garages that repaired and serviced this make of automobile would be a quick and inexpensive way of collecting data about which radio station the business should spend its advertising budget with. It could also be used to verify claims by radio stations about what segment of the market they reach.

This method could also be used to canvas magazine or newspaper readership preferences by scanning the titles of these types of publications that are discarded by residents of a target neighborhood on the day paper products are left at the curbside for recycling.

One consideration that needs to be kept in mind when using this technique is that because data are being collected without the target's consent, there may be ethical and legal issues to consider. For instance, with the radio station collection example, one would need to consider whether checking the car's radio without the owner's permission would violate any state or federal statutes regarding privacy. With the example of inspecting discarded newspapers or magazines, it is unlikely that this data collection would be deemed either a trespass or an invasion of privacy by a court as these goods, by virtue of their abandonment, are no longer the property of (at that stage) the former owner.[4] However, there may be local bylaws that regulate interfering with material placed on the curbside for rubbish collection or recycling. So, before incorporating unobtrusive methods into an information collection plan, seek legal advice.

More often than not, indirect methods will be appropriate because the data sought are readily available. *Reliability* is an estimate of the consistency a collection instrument will yield each time employed or the *consistency* between two or more sets of data that have been collected. Using this technique is also likely to be ethical as well as legal. But just as with other data collection methods, the analyst will want to assure themselves of the reliability of the collection method. In this regard, indirect data collection techniques can act as a check for the reliability of another data collection method that may be being used (e.g., a direct method) or for estimating the reliability of itself via a test-retest approach. This is done by collecting data at two different points in time and computing the correlation between the two sets of data. If the analyst considers that there is no change in the underlying condition between the two tests, then the data's reliability can be estimated.

CONTENT ANALYSIS

Content analysis is the analysis of text contained in documents. Because these data are usually in narrative form (e.g., transcribed oral speeches, media interviews, or open letters to a public audience), the most common form of analysis is via qualitative methods, but analysis can also incorporate quantitative methods or both. The central purpose of content analysis is to develop an understanding of what is contained in the text beyond the superficial message. Some commonly used techniques for content analysis include thematic analysis, indexing, and qualitative descriptive analysis.

More specialized techniques, such as psychohistorical and psycholinguistic analysis, can also be used to provide clues about the target's probable course of action. These techniques are based on an examination of behavior patterns exhibited by the target (which can include a nation state or other actor) in previous written (or transcribed oral) communications. The assumption is that behavior patterns are manifested in these communication forms, and through a psychoanalytic profile of the target, the analyst can distill a set of behavioral indicators. These profiles provide insight for basing recommendations for action. This technique requires the analyst to have expert subject knowledge (e.g., clinical psychology or psychiatry) to credibly assess the data. If not, help can be enlisted from private practitioners and university academics, depending on the security classification of the project.

On October 6, 1973, Israel suffered a surprise attack launched by Egypt and Syria—the Yom Kippur War. Israel failed to recognize or misinterpreted indicators that could have provided the insight it needed to stand up troops in preparation for war. If a psycholinguistic analysis was carried out of the provocative speeches by President Sadat weeks before, it may have revealed such indicators.

Similarly, psychohistorical analysis is based on the assumption that a country's behavior may be influenced by a range of factors that make up its cultural identity, including social, anthropological, political, and noteworthy historical events. For example, a psychohistorical analysis of Latin America would need to consider the issue of U.S.–Latin American relations. This is because U.S. interests in Latin America go back centuries, with the U.S. intervening in many of the region's countries over that time. As such, an evaluation of the possible ramifications of any actions would need to be done as a number of those countries have felt a high degree of anti-U.S. sentiment. The benefit of such a psychohistorical analysis could provide the insight needed to avoid the political criticisms experienced after the 1961 Bay of Pigs incident.

Thematic analysis (a form of content analysis) is where a set of themes is distilled from the text. Ideas that present themselves in the reading are identified and given appropriate labels. The passages of text relating to a particular theme are marked and affiliated to the corresponding theme. Using a manual system, this could mean photocopying the page text; cutting out the word, passage, or paragraph (using a pair of scissors); and placing it in a large envelope along with other passages that are identified. Each piece of paper (containing a word, passage, or paragraph) is referenced back to the text as one would do using, say, the Harvard referencing style so that when the material is compiled into the final report, the analyst knows where it has come from (similar to note taking when writing an essay).

If a computer-based software package is used, then the same procedure takes place, but the software avoids all of the manual cutting, referencing, and storage issues as this is done electronically. The advantages of using a computer-based solution are great as the software will have other features, including word count and metrics such as the Flesch Reading Ease and the Flesch-Kincaid Grade Level tests (see the section later in this chapter). These types of analysis allow the analyst to use a blended approach, incorporating both qualitative and quantitative analysis.

Any number of documents can be analyzed using this technique, and in cases where, say, speeches are delivered by a foreign political figure in the country's local press, these public addresses can be analyzed over time in a quasi-longitudinal study. Alternatively, a number of author-known documents can be analyzed against a document of unknown or uncertain authorship with this tool.

As an illustration, the use of the Flesch Reading Ease and/or the Flesch-Kincaid Grade Level tests could be used where an analyst wants to gauge whether a speech was aimed at the local population or whether its appearance in the daily newspaper was just a vehicle for projecting a message to international leaders. The use of these two tests could quickly yield results that suggest whether the speech under investigation had a reading ease score and grade level commensurate with the country's population or that it was much higher, perhaps suggesting it was aimed at a better educated international audience.

Indexing is another way of identifying meaning in the text. Rather than identifying themes, indexing identifies keywords in context. This is best done with computer software. An exceptions dictionary is first set up in the software package where words such as *a, an, and, the, is, it, of,* and so on are flagged as exceptions, so when the software searches the text, it does not index these inconsequential words. All other words are then indexed.

These words appear in their context within the document. This allows the analyst to not only count the appearance of certain words—say, a repeated word like "infidel"—but also then tag the sentence or paragraph where these key words appear so that they can be included in further analysis of a wider theme. Indexing is a technique closely tied to quantitative descriptive analysis. Descriptive analysis seeks to describe features of the text quantitatively as is done with numeric data—by describing the most frequently used words or phrases. This type of analysis ideally lends itself to a blended approach of both qualitative and quantitative techniques.

Although content analysis has the advantage of being unobtrusive, and computer-based tools allow the analysis of very large documents (and multiple documents), the technique is not without limitations. Firstly, the data need to be textual. If there is no source of textual data, no analysis can take place.

Caution must be exercised in terms of sampling bias as with other approaches. In the hypothetical case of the foreign leader's speeches in the local press discussed above, analysis would leave out all of the leader's speeches delivered by way of radio or television broadcast or in-person delivery to crowds assembled. The other limitation is that although software packages are useful in automating indexing, counting, and tagging text, these packages cannot interpret what words or phrases mean—this takes an analyst to do this word by word and phrase by phrase.

Flesch Reading Ease

The product of the Flesch Reading Ease analysis is a rating represented by a number on a scale between 0 and 100. The higher the score, the easier the text is to read. For instance, a document with the following Flesch Reading Ease scores would be interpreted as such:

- 90 to 100—very easy;
- 80 to 89—easy;
- 70 to 79—fairly easy;
- 60 to 69—considered to be what is generally termed *plain English*;
- 50 to 59—fairly difficult;
- 30 to 49—difficult; and
- 0 to 29—confusing.

The formula for calculating the score is: *readability ease* = 206.835 − (1.015 × *average sentence length*) − (84.6 × *average syllables per word*). Although most word processing packages will have this feature as part of its spelling and/grammar checker, it will be of interest to the intelligence analyst to understand how it is calculated. The average sentence length is determined by taking the total number of words in the document and dividing it by the number of sentences. The average number of syllables per word is calculated by taking the total number of syllables divided by the total number of words. Step by step:

1. Count all the words;
2. Count all the syllables;
3. Count all the sentences;
4. Calculate the average number of syllables per word;

5. Calculate the average number of words per sentence;
6. Match the readability score.

Flesch-Kincaid Grade Level

The Flesch-Kincaid Grade Level analysis converts the Flesch Reading Ease score to a level equivalent to grade school rank (U.S.-based). The formula for calculating the score is: *Flesch-Kincaid Grade Level* = (0.39 × *average sentence length*) + (11.8 × *average syllables per word*) − 15.59. For example, a score of 12 indicates that a person who has had a twelfth grade education could understand the text contained in the document. Step by step:

1. Count all the words;
2. Count all the syllables;
3. Count all the sentences;
4. Calculate the average number of syllables per word;
5. Calculate the average number of words per sentence;
6. Multiply the average number of words per sentence by 0.39, and add this number to the average number of syllables per word, which is multiplied by 11.8;
7. Subtract 15.59 from the resulting number from step 6;
8. Match the score with a U.S.-based school grade.

If a word processor or other software package is used to perform these calculations, note any limitations specified in the software as some results may only report a grade level of 12 even though the grade level exceeds this figure.

SECONDARY ANALYSIS OF DATA

Secondary analysis of data in a quantitative approach is like its close associate—content analysis, which relies on data that have been collected already. But rather than analyzing textual data, secondary analysis analyzes quantitative data for a second time; this examination is unconnected to the primary collection effort.

By way of example, intelligence analysts used secondary data analysis in the lead-up to the 1973 elections in France. At the time, decision makers in the United States were interested in knowing if a socialist-communist-left radical coalition was likely to form a government in France. Therefore, U.S. decision makers requested an intelligence assessment on the most likely outcome of the election. Intelligence analysts at the Central Intelligence

Agency (CIA) were tasked with providing the assessment. Relying on existing data sets, CIA analysts used multiple regression analysis to gauge the impact various historical economic conditions had on the voting patterns of the left.[5]

Based on the results of this secondary analysis, analysts concluded that economic conditions did, in fact, impact elections but only in the absence of other important political considerations.[6] The use of existing data sets allowed analysts to predict that the domestic political factors affecting the elections would not be as powerful as those in previous elections—therefore raising the potential of a left victory. However, these political factors were also assessed as being strong enough to ensure that the required number of votes would *not* go to a left coalition. Arguably, such analysis was only made possible by using this unobtrusive technique.

The amount of data collected by governments around the world for social and economic planning is extensive. There are census data, crime statistics, social data, educational data, economic data, and consumer data, just to mention a few. Data are also collected by private corporations, think tanks, and a variety of nongovernment organizations for their own planning purposes, which are also available to outside researchers. Because these data are stored electronically, they can be imported into the software package being used by the analyst—say, a spreadsheet or a statistical package such as SPSS. As was seen in the 1973 French election example, CIA analysts used data from several databases to conduct their multiple regression analysis.

Secondary data analysis is an efficient means of conducting research. Preexisting data sets alleviate the problem of a potentially lengthy collection phase that, in some cases, may have taken months or years to collect. It also means the data may be available at no cost or a modest fee, thus making it inexpensive as well. If a pretext is used to obtain the data set, it will not alert the target (or target country or corporation) that they are the subject of an intelligence operation. For instance, most small intelligence research studies are unlikely to have a budget large enough to conduct a national sample (or international as in the French case cited) because of the cost and time required, but by using a census data set, even a small research budget can gain considerable leverage.

Secondary analysis does have some limitations. In the main, analysts may find it difficult to gain a full appreciation of the problems encountered during the original collection or the errors inherent in the resulting data sets so that these limitations can be taken into account when manipulating the data during secondary analysis. It may also be a difficult task to link two or more data sets that have little in common—either by the structure of the database, the level of measurement (i.e., nominal, categorical, interval, or ratio), or units of measurement.

> "Not everything that counts can be counted, and not everything that can be counted counts."
>
> —Albert Einstein

KEY WORDS AND PHRASES

The key words and phrases associated with this chapter are:

- Consistency;
- Content analysis;
- Flesch-Kincaid Grade Level analysis;
- Flesch Reading Ease analysis;
- Indirect data collection;
- Open source information;
- Psychoanalysis;
- Psychohistorical analysis;
- Psycholinguistic analysis;
- Reliability;
- Secondary analysis; and
- Unobtrusive methods.

STUDY QUESTIONS

1. Explain why unobtrusive data collection methods are considered intrinsically safe.
2. Cite three unobtrusive information gathering techniques.
3. Discuss one way an automated system for collecting indirect data can be set up using the Internet.
4. Outline the main advantages of using unobtrusive methods.
5. Explain how quantitative analysis can be conducted with unstructured data, say, in the form of a leader of a country's public speeches.
6. List five sources of secondary data.

LEARNING ACTIVITY

Suppose that Country Q has had a sudden change of its leadership regime. As such, the agents who were in place previously have lost access to all secret sources of data. Consider how you as an analyst might use unobtrusive data collection methods in the short term to provide the intelligence needed to monitor the situation until field operatives reestablish their agent network(s).

NOTES

1. Harry Howe Ransom, *The Intelligence Establishment* (Cambridge, MA: Harvard University Press, 1971), 19. Professor Ransom was quoting Ellis M. Zacharias, a World War II deputy director of the Office of Naval Intelligence. According to Zacharias, only 4 percent of intelligence came from semi-open sources, and a mere 1 percent from secret agents.

2. Ransom, *The Intelligence Establishment*, 20.

3. See, for example: Henry Prunckun, "It's Your Money They're After: Sting Operations in Consumer Fraud Investigation," *Police Studies* 11, no. 4 (Winter 1988): 190–94; Henry Prunckun, "Sting Operations in Consumer Fraud Investigation," *Journal of California Law Enforcement* 23, no. 1 (1989): 27–32.

4. Rick Sarre and Tim Prenzler, *The Law of Private Security in Australia*, 2nd edition (Pyrmont, Australia: Thomson Lawbook, 2009).

5. Susan Koch and Fred Grupp, "Regression Analysis: Impact of Economic Conditions on Left Voting in France," in *Quantitative Approaches to Political Intelligence: The CIA Experience*, ed. Richards J. Heuer (Boulder, CO: Westview Press, 1978).

6. Koch and Grupp, "Regression Analysis," 57.

6

✺

Covert Sources of Information

This topic examines secret sources of information in the context of the wider spectrum of data available, including information that can be obtained from:

1. Open sources;
2. Semi-open sources;
3. Clandestine sources; and
4. Covert sources.

INTRODUCTION

The spectrum of information sources includes open, semi-open, clandestine, and covert. This chapter examines covert sources of data that are available to the intelligence analyst. Because of their intrinsically safe nature, open and semi-open sources of information do not require the same high level of planning and consideration when compared to organizing data to be collected by clandestine and covert means.

Open and semi-open sources can be dealt with in an up-front way with the analyst's approaching the source and making a request for the data. This includes approaches such as interviews, surveys, questionnaires, and other forms of direct data collection. These particular methods are adequately covered in the literature relating to applied social science research, so analysts are referred to this body of information. These topics are only touched on here.

Clandestine data collection, although akin to covert collection, is different in that clandestine operations operate in the open—visible to the target but disguised so that they do not appear to be what they seem. Covert operations, in contrast, are carried out in secret. They are hidden and not visible to the target even in a disguised form; they are to some degree intrusive, but because they are invisible, the target has no knowledge that an operation is being conducted.

These methods are discussed here and include: undercover agents, physical surveillance, electronic surveillance, informants, mail covers, and waste collection. Analysts may want to use these methods in their collection plans because an attempt to obtain information via open and semi-open methods might be met with a nil result. For instance, covert methods are used where the target is concealing information. The only way to obtain such data under these circumstances might be to penetrate the security measures via covert methods.

OPEN AND SEMI-OPEN SOURCES

Because we live in an age where information is central to every aspect of life, an astonishing variety of obligations have been placed on individuals to recorded information about their affairs. The same applies to corporate bodies and governments. It is because of these record systems that society generates what has become known as a *paper trail*.

A paper trail can be described as all records and documents created by an individual or entity in the course of commercial and social interaction with other individuals, organizations, government departments, and businesses (both public and private). These records and documents have the effect of leaving a trail detailing where the individual (or entity) has been, with whom they have had dealings, what goods and services they have purchased, what they own, what their likes and dislikes are, and more importantly, what their intentions may be.

To the analyst, this trail forms a composite picture for any target that comes under surveillance. Uncovering one part of the paper trail can lead the analyst to other sources of information. These sources are not only limited to those on paper but also can be extended to interviews with friends, neighbors, and colleagues of the target, perhaps using a pretext as well as physical, optical, or electronic surveillance.

To the analyst, the paper trail is a valuable set of leads. The value of this information in a collated and analyzed form can be seen in the arrests reported in the media from time to time of influential organized crime members that were, until that point, untouchable by law enforcement agencies.

Sources of open and semi-open information are by no means limited. Open source information (abbreviated OSINT for open source intelligence) can be obtained from newspapers, magazines, professional journals, radio and television broadcasts, and the Internet. Arguably, until the terrorist attacks of 9/11, intelligence analysts used these sources of information simply as a means of supplementing classified information. The importance of systematically collecting and analyzing open source information did not become a priority for the intelligence community until after these attacks. For instance, the effective mining of Internet-based information has enabled the intelligence community to better understand how jihadists use the Internet's web-television capabilities, chat rooms, and news sites to train their members and raise money.[1]

Figure 6.1. Some of the texts on secret intelligence held by the Russell J. Bowen Collection of Works on Intelligence, Security, and Covert Activities at Georgetown University.

Examples of more traditional sources of information include but are not limited to: telephone directories (both current and backdated); city directories; vehicle license plates; drivers' licenses; birth, death, and marriage records; civil and criminal court records; property titles, mortgage documents, liens, and caveats; school records; voter registration lists; credit reporting agencies; utility companies; credit card companies; insurance companies; stockbrokers; moving companies; chambers of commerce; racing or gaming commissions; banks and finance companies; the postal authority; numerous government departments, agencies, and statutory bodies (local, state, and national); and employment agencies.

A sometimes overlooked source of information is the public and university library. In addition to the wealth of information contained in the library's stacks, libraries have reference books, maps, newspapers, journals, periodicals, registers, and catalogs. There are also special collections within libraries and privately maintained computer databases that contain topics of particular interest. An example of the former is the Russell J. Bowen Collection of Works on Intelligence, Security, and Covert Activities, which is housed in the Lauinger Library, Georgetown University, Washington, DC (see fig. 6.1).

The United States has its government printer, the Library of Congress, Congressional Record, National Archives, and information available through the Freedom of Information Act. In most liberal democracies, there exist equivalents to the freedom of information legislation as well as equivalents to the holdings of the U.S. archives cited. Finally, there are public radio and television broadcasts, which have the potential of providing an extremely wide range of information, as do photographic and motion picture archives.

UNDERCOVER AGENTS

> The objective of the undercover agent is to infiltrate "as deep as possible and [gather information] on the opposition or enemy."[2]

Undercover agents are able to get close to individuals or inside organizations to make firsthand observations. The use of agents is risky for the agents and the organization for which they are employed. These risks are both physical and psychological.

The agent risks physical harm in the form of bodily injury and death as well as a range of psychological injuries spanning from mild anxiety to severe psychiatric disorders. The physical risks are more apparent as one can easily visualize the ramifications of having to penetrate an illegal enterprise. The psychological injuries arise from the stresses associated with working in isolation, working in a dangerous environment, and perhaps engaging in activity that is illegal (including consuming illicit drugs and alcohol in binge quantities).

The data that can be obtained from an undercover agent can be very valuable because it is an opportunity to get a glimpse of the target's intentions, thereby providing an insight into the target's thinking, rationale, and behaviors that could not be obtained by other means. However, given the risks and the monetary costs of "running" an agent, it is a method that is not often used in the first instance. It is usually reserved as a means of last resort or for targets that pose a serious danger to society or national security.

Because the agent will be in direct contact with the target, there are a number of issues that must be kept in mind. The most important is that the agent's identity must be guarded with utmost secrecy. If knowledge of the agent leaks to the target, the agent's cover will not only be "blown," but also the agent is likely to suffer injury.

Part of the agent's brief is to obtain evidence of wrongdoing but also intelligence data. Evidence may be in the form of admissions, but intelligence data may be in the form of indicators of intent. To capture these data, the agent can either commit the details to memory and then record them later for transmission back to the analyst's agency or use some electronic device that transmits the data live for recording and transcription. The latter is the most reliable and the best solution as it doesn't rely on the agent's ability to remember the details, which, from an intelligence point of view, can be critical. Data can be qualitative (e.g., discussions with the target) and quantitative (e.g., numbers of items, times, routes, colors, preferences, and so on).

INFORMANTS

Using informants to gather information places a safe distance between agency's operatives and the target. An informant is someone who has indicated willingness to assist the agency in achieving its goals—whether that is to arrest a drug trafficker, close down a trademark infringement ring, or sell state secrets to a friendly government. Informants do these things for a variety of reasons. As T. J. Waters discussed in his memoirs *Class 11: Inside the CIA's First Post-9/11 Spy Class*:

> Values + beliefs = behavior. This is what you need to keep in mind. Know your target's motivations in the context of their values and beliefs. Values

govern behavior as motivation for our actions. Beliefs are how we express our values to ourselves and others. . . . We must understand the agent's history, the background fundamentals of how they became who they are in order to understand their motivations.[3]

The processes for informant handling vary from agency to agency and from informant to informant, but in general, the process of employing an informant starts with some form of "registration." This serves several purposes, the main being accountability of the funds (or other gratuities) that will ultimately be paid to the informant for service.

Corruption is a temptation when running informants because the system relies on the honesty of the informant handlers to pass over the full amount of funds without "skimming" any for themselves. Also, there is the temptation to run fictitious informants in order to collect payment; this ruse was made legendary by Graham Greene in his book *Our Man in Havana*, whose main character, Jim Wormold, ran an entire spy network that did not exist but regularly passed on fictitious intelligence reports.[4]

As such, informants are photographed, fingerprinted, and their vital statistics recorded on an official register. This register is maintained by the agency, and the data it contains are classified so only those that have a need-to-know can access it. This helps to protect the informant's identity and, therefore, safety.

The data that informants can provide will vary greatly—from accurate and reliable down to inaccurate and unreliable. If the informants are not trained in observation skills or memory retention techniques, the data they will provide may be sketchy. If however, they are trained—as in the case of a foreign intelligence operative who has "swapped sides"—the data may be very sound. To overcome training shortcomings, a listening device may be employed (such as a body transmitter), but this option may be reserved to where the informant is only used once or twice as the risk of discovery and personal harm is great—with each use, the chance of detection increases. Also, there has to be an agency listening post nearby to receive the transmitter's signals and record the audio. Having a van or other vehicle containing the listening post equipment parked near the target's location may unduly raise suspicion.

If an apartment is rented close by to set up a more permanent listening post, this will incur added financial costs and open up another possible source of leaks—for instance the realtor or landlord may "talk" if the agency has rented the property openly without using a cover. If the agency has rented the premises under a cover, the foot traffic in and out of the building may expose the operation. As with undercover agents, the use of an informant to gather data must be weighed against a number of risks: exposure of the operation, harm to the informant, the likelihood of obtaining the information sought, and the likelihood of obtaining that information accurately.

In regard to the last point, it will always be the case that the analyst needs to try to verify the accuracy and reliability of the informant's data. There may be many ways of doing this, but one example is to compare the data against similar information provided by the same source. Whatever way it is done, it should be borne in mind that there is some degree of uncertainty inherent in running informants. Their motivation can be greed, revenge, or exchange for some favor (e.g., reduced jail sentence), anything *but* the pursuit of truth.

PHYSICAL SURVEILLANCE

> Operatives "assigned to a surveillance operation should possess a reasonable amount of resourcefulness and adaptability so that [they] can effectively blend with the environment, both in appearance and conduct."[5]

Physical surveillance is the act of making observations of people, vehicles, or the activities occurring at specific locations. Many research methodologies use observation as the key means of collecting data. This can take the form of a researcher standing on a street corner and counting pedestrians as they pass by or as they enter a building, or it could be a data collector who records the number of vehicles that pass a particular intersection or travel along a suburban street. Surveillance used in intelligence operations is usually covert and takes the form of either moving surveillance or fixed surveillance.

Physical surveillance may take place either at a particular location, which is known as a *stakeout*, or in a continually moving situation, referred to as a *tail*. Physical surveillance is often used to supplement information which has been obtained from open or semi-open sources. It can also be used early in an operation to accelerate the generation of leads, corroborate existing information, or obtain details that would not be available through other avenues of inquiry.

Fixed surveillance is a tedious and time consuming activity, but its value should never be underestimated. Its strength lies with the field operative, who is tasked with a single mission—to obtain details about the target. The results can be pivotal to any intelligence research project. Unlike social research where a data collector stands in the open with a clipboard and counts passersby, the covert surveillance officer, termed an *operative*, may sit for hours in the darkened interior of a van or in the backseat of a car, video recording the comings and goings of the target and associates.

Moving surveillance is when the operative follows the target. The objectives of the moving surveillance are the same as a fixed surveillance but, in addition, tailing the target wherever they go. The most common form of moving surveillance is via car as this is the usual mode of transportation people use. Surveillance on public transport—aircraft, bus, train, and tram—is also possible, but the level of risk of detection increases as following a target in such close proximity makes the operative's task more difficult.

A benefit of covert surveillance is to validate the information coming from undercover agents and informants. For instance, a surveillance operative should be able to independently verify that, say, an informant did meet with the target on the day and time provided by an informant. Although it is highly unlikely that the operative will be in a position to verify what was said between the parties, validating the meeting may go a long way in establishing an informant's creditability, especially if the pattern is repeated. However, there may be other ways of validating the details of a meeting if a surveillance operative is able to video the meeting.

For instance, the length of the meeting may be able to suggest whether the details the informant has purported during the conversation—a two minute conversation on a street corner outside a coffee shop is not consistent with details of a long planning meeting over a lengthy meal at a restaurant. Surveillance would be able to shed light on this. Likewise, the body language of the two as they talk may be an indicator of how well the meeting went—was there aggravation, frustration, disbelief, threats, or did it conclude with mutual admiration, gratitude, and a relaxed atmosphere? Such independent observations could prove valuable in supplementing or validating other covertly obtained data. The important point is that surveillance is actual observations, not self-reported information; the latter can be subject to any number of biases.

OPTICAL SURVEILLANCE

Because operatives rely so heavily on their eyesight in conducting physical surveillance, various devices are used to assist them to extend their vision. Using technology also assists operatives to position themselves much further away from the target than could normally be achieved with unaided sight, thus reducing the possibility of discovery (see fig. 6.2).

The principal and traditional device used in physical surveillance is a pair of binoculars. Binoculars are an optical device consisting of two prism-operated telescopes fixed in parallel. This configuration enables an operative (sometime termed the *surveillant*) to view a magnified image of the subject using both eyes. Binoculars have been designed to provide both magnifying power and light gathering capabilities. The latter is essential for surveillance work at dusk and at night. Binoculars that have a built-in digital camera are also available commercially.

Viewing a subject at distances that exceed the effective range of binoculars is accomplished by using a telescope. The magnifying power of the telescope ranges from 20 to several hundred times that of normal vision. Binoculars, in comparison, range from about 6 to 20 times that of normal vision.

A viewing device usually at home in submarines, but sometimes used in land-based information gathering, is the periscope. Small, portable high quality devices are used to peer into high, inaccessible windows, over walls, and around corners. Periscopes are also used in applications such as surveillance vans (e.g., mounted on the roof).

Recent developments in optical technology have now made low light (night) viewers available at affordable prices. These units are designated as either "active" or "passive" night vision devices. The first group, the active devices, are devices that operate by using an infrared light beam. The surveillant projects the invisible beam of energy so that it illuminates the subject. The image is then viewed with special equipment, which converts the infrared radiation into the visible light spectrum. The second range of devices operates by amplifying the existing background light—moon, stars, street lights, and so on—by several thousand times, thus literally turning night into day through the sights of the surveillant's night scope.

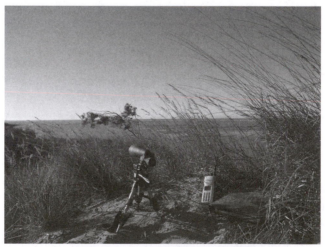

Figure 6.2. Covert optical surveillance of illegal fishing—Port Roper near the Gulf of Carpentaria, Australia (photograph by Intelligence Officer Nathan McGrath of the Northern Territory Police, Fire, and Emergency Services, Australia).

COVERT PHOTOGRAPHY

By far, the most commonly used camera in covert photography is a digital single lens reflex (SLR) (which replaced the older 35 millimeter film-based camera). However, in some unique situations, such as surreptitious copying of documents, a subminiature camera would usually be employed. The value of subminiature cameras lies in their concealable size (e.g., mobile/cell phones with built-in cameras). Nevertheless, the advantage of digital SLRs are their fast (light sensitive) lenses. The ability to download images, edit them using commercial software, and send them over the Internet applies to all digital photography.

As with binoculars and telescopes in physical observation, telephoto lenses play an important role in covert photography. A reflective mirror lens (catadioptric) not only enables a surveillant to greatly reduce the physical size of his equipment, making concealment easier, it acts to extend the agent's operational distances. Basically, catadioptric lenses use a system of mirrors to compress the light's optical path. Magnification (measured in the focal length of the lens) varies from about 100 to 2,000 millimeters.

Digital video cameras are also popular for surveillance work. These cameras are no larger than SLRs—and, in a growing number of models, smaller—so they are often used with a telephoto lens, thus allowing surveillants to record their observations from safe distances. In the case of a fixed position surveillance (a stakeout), a time-lapse option can be used to provide extended coverage from a single digital memory device. In order to achieve this, time-lapse photography operates in a frame-by-frame mode at greatly reduced memory (25 frames per second is considered real-time recording). Many state-of-the-art video cameras can be built into such items as briefcases, books, wall clocks, paintings, and plants.

AERIAL PHOTOGRAPHY

Aerial photography from both rotary and fixed wing aircraft is another effective method of information gathering. The history of aerial surveillance dates back to the mid-nineteenth century when the first photograph was taken from a French military (hot air) balloon.

Since that time, many developments have occurred in the area of photoreconnaissance. The technology currently available ranges from reconnaissance satellites and ultra-sophisticated spy planes down to the light plane-for-hire with conventional handheld photographic equipment. The former, of course, are used by intelligence agencies of various nations, while the latter would probably be used on a rental basis by smaller agencies on an ad hoc basis.

Internet-based mapping facilities give analysts an easy and cost-effective option for obtaining aerial photographs. Websites provide satellite images of the earth and cover almost all locations on the globe. Intelligence analysts can select several different types of views: satellite imagery, maps, and terrain, or combinations. Street-level views are also available. Although these sources would not be practical for the analyst who is monitoring an arms control agreement (for the most part, these are static images refreshed only occasionally), they may be all that is needed to brief a local police or private investigation surveillance team as to, say, reconnoiter the terrain surrounding the target's place of business.

ELECTRONIC SURVEILLANCE

Electronic surveillance is most likely to be in the form of audio surveillance (i.e., room listening devices or *bugs*), telephonic intercepts, or Internet-based intercepts. There are other forms of electronic surveillance, such as radio frequency intercepts (i.e., the interception of radio transmissions), but the most commonly used intercepts are from these three methods.

Audio Surveillance

The fundamental principle of any audio surveillance operation is to be able to plant a quality microphone as near as possible to the target and, in doing so, avoid background noise that may render the intercept unintelligible. The types of microphones used in audio surveillance are required by the nature of the work to be very small and are referred to in the trade as *subminiature*. Their minute size allows them to be secretly deployed in the target's environment. Once in place, they can be connected to a high gain amplifier, then to headphones (for live monitoring), a digital recorder (for listening at a later time), or a transmitter to relay the audio to a listening post.

If a transmitter is not used to send the intercepted audio, then wires are needed; the wires leading from the microphone to the listening/amplifying equipment must be concealed to prevent detection. In some cases, existing wires such as telephone, electrical, or even, paradoxically, burglar alarm wires can be used instead of running new wires, thus making detection more difficult. There are also special metal paints on the market which an operative can use to "paint" wires across a room. Once touched up with paint of the surrounding decor, these "wires" are reported to be very difficult to detect (especially if used in an industrial or warehouse setting).

There are several types of microphones the surveillant can use depending on the particular situation. They are the tube, contact/spike, pneumatic, and directional microphones.

Tube microphones are designed to be inserted into a targeted room via a very small drill hole. This species of microphone consists of an element connected to a thin tube. The tube emerges flush with the wall in the targeted room and can only be detected by very close inspection. Tube microphones are likely to be located in spots made classic in spy novels— behind a wall, a picture, a piece of furniture, anything that would hinder detection for as long as possible.

Contact and spike microphones, in contrast, require less effort to secure their installation. These microphones do not respond to air vibration as a conventional microphone does but rather translate vibration into sound. Contact microphones are similar to the "pick-ups" used by musicians to amplify their instruments, and spike microphones resemble the vintage phonograph needle. These types of microphones are attached to the exterior of a wall, window, floor, or ceiling. Once in place, these devices will reproduce quite clearly the sounds produced within the targeted room. These microphones lend themselves well to permanent installation.

The pneumatic cavity microphone is the electronic version of the drinking glass placed against the wall trick, historically recognized as an effective method for monitoring adjacent room conversations. The pneumatic microphone is substantially superior and operates by using a specially constructed shell which is highly responsive to surface vibrations at audio frequencies found in the range of human speech. This "cavity" is used in conjunction with a conventional microphone element to enhance the

Figure 6.3. Drinking glass placed against the wall trick.

device's performance. It also forces audio output to correspond to a wall's surface (or window, floor, or ceiling, depending on the case) vibrations rather than a direct sound output.

Directional microphones are of two types: parabolic and shotgun. Parabolic microphones consist of a "dish" with an inwardly pointing microphone element. The targeted audio is reflected and focused into the microphone element, thus gaining a directional effect. The shotgun microphone (also known as a *rifle* or *machine gun* mic) operates on the same principle but utilizes a long tube, or set of tubes, in a cluster to pinpoint the targeted conversation. The effective range of these microphones is reported to be about 150 meters, and they are known to be able to pick up audio through closed windows at closer distances.

> Radio transmitters offer an agent much greater degree of safety from detection because installation requires far less effort than a hardwired device.

Another area of audio surveillance, discussed briefly above, is that of wireless microphones, also known as miniature radio transmitters. These devices rank further up the hierarchy in surveillance sophistication. These devices do not have any wires or connections to reveal their location. They can be attached to furniture or fixtures by means of a magnet or adhesive surface or concealed in everyday objects such as rings, pens, cigarette lighters, books, ashtrays, or pictures. Transmitters do not require the eavesdropper, or his equipment, to be located nearby. Their range of transmission varies from 60 meters to almost a kilometer. It is directly dependent upon transmitter strength, the thickness of surrounding walls, the sensitivity and selectivity of the receiver, as well as its antenna system.

Body transmitters are generally larger, more powerful, and better constructed than wireless microphones. This is because the chance of detection is slight, and the device is intended to be used repeatedly. These devices are designed to operate from an operative's coat pocket, underclothing, or attached by tape directly to the agent's body.

Akin to the body transmitter is the briefcase transmitter. Not only can this device be used by a "walk-in spy," but it can be conveniently "forgotten" in the target's room or office in order to obtain ensuing conversations.

Another type of electronic surveillance transmitter operates by broadcasting in the very low frequency range (VLF is between 3 kHz and 30 kHz). This device uses electrical power lines for signal transmission. The signals move along the wire path, and, because of the device's very low frequency, very little energy is radiated into space. This method of communication is used by many of the household wireless intercoms sold commercially.

Communication equipment which operates outside of the standard FM radio broadcast frequencies tends to be more secure from interception (that is, outside of the 88 MHz to 108 MHz frequency range). This is because the radio receivers needed for this type of radio reception are not sold in regular retail outlets. Nevertheless, they are available commercially at radio and electronic stores and are a common feature in the radio rooms of radio amateurs (i.e., "ham radio"). There is also a cadre of radio enthusiasts known as listeners who scan the radio spectrum listening for any unusual content or sources. They have high-gain antennas and receivers that scan many hundreds of individual radio frequencies each second and, as such, could detect the conversation collected by a bug. If there is a chance of this, the radio technician who will oversee the bug installation should consider the use of a scrambler or other form of encryption.

Some receivers operate in the microwave part of the radio spectrum, such as those used by telephone companies for telecommunications and by private enterprises for computer data transmission. Such receivers are complicated in design and expensive; although, military and governmental intelligence agencies would have ready access to these. Units used for intercepting microwave communications can be set up in a van or building anywhere along the path between the transmitter and the receiver, which could be several hundred kilometers long.

Several devices, although not specifically designed for surveillance work, are worthy of a brief note because their function provides the operatives with an increased scope of application. These are the drop-out relay, the voice-operated relay (VOX), and the carrier switch.

The drop-out relay is attached to the target's telephone line and then wired to a transmitter. It will switch the bug on whenever the hand piece is lifted from its cradle and off when it is replaced. This prolongs battery life, and because the bug is not transmitting continuously, it lowers the risk of being detected by an electronic countermeasures sweep.

A VOX is similar in purpose to the drop-out relay. If connected to a room transmitter, the bug remains dormant until activated by the sound of a voice or noise. The VOX turns the bug on and off as individuals enter and leave the targeted room.

Finally, a carrier switch can be used to start and stop a digital recorder when it receives an audio signal, and it can be activated by a hidden transmitter. Employed in conjunction with a drop-out relay or a VOX, the switch carrier increases exponentially the surveillance capabilities of the operative.

Telephone Intercepts

The telephone is a very useful medium for electronic surveillance. Obtaining information using the telephone involves two methods: the first uses devices that intercept conversation directly from landlines and requires no entry into the target's property, and the second is one that uses a portion of the telephone system for room eavesdropping and usually requires physical access to some part of the telephone system.

Wiretaps

Wiretapping is the interception of fixed line telephone and facsimile communication, as well as computer data that are transmitted over landlines. The interception of these signals can take place anywhere between the target's location and those of the surveillant. The more difficult parts of the telephone system to install a device are at the target's property and the lines leading out of that building. Once the targeted line(s) leaves the building and melds with the wider telephone network, interception is far less difficult because with a court ordered warrant, a surveillant can install unobtrusive equipment with greater ease.

Because the telephone company provides all of the electrical power required to operate a target's telephone service, this medium offers great eavesdropping benefits to operatives. For example, the telephone's power supply can be used to directly operate electronic eavesdropping devices; the wires themselves can be used to carry the resulting audio signals, and the microphone in the handset can be used to listen in on room conversations.

The techniques used to tap conversation from telephone lines consist of: (a) direct wire connections and (b) induction coils. In direct wire connections, the lines are cut, and the listening device spliced in place, using an electronic matching network. With induction coils, the tapping process literally lifts the audio signals off the telephone line and, therefore, does not require splicing. For this reason, induction coils can prove to be undetectable by electronic countermeasure sweeps. The only effective way to detect this type of bug is by visual inspection of the telephone wiring.

An alternative to direct wire connections is a radio system. This is the same as the wireless microphones described previously, except that it requires no microphone element. This is because the audio signals are already in an electrical form. Radio systems use the telephone line voltage for power, as opposed to batteries, and transmit whatever conversations are on the targeted line to a remote receiver/digital recorder.

The placement of telephone surveillance devices can be quite arbitrary; the only limit to their deployment would be the ability of an operative to gain access to the target's telephone system. The devices may be installed within the telephone instrument itself, anywhere along the line in the targeted building, on a telephone pole outside the building, or in the wire closet or terminal room where the lines are joined to the branch feeder cable.

There are several other electronic telephone surveillance devices which allow an operative to record the telephone numbers and dates when dialed by a target but do not permit the recording of the conversation. These are the dial impulse recorder, commonly referred to as a pen register, and the touch tone decoder. They operate simply by counting the impulses in each dial pulse group, that is, the digit dialed. An alternative to these two devices is the variable speed tape recorder. This operates by recording the desired conversation, then replaying it at a reduced speed so that the impulses can be counted to determine the number dialed.

The harmonica bug, or infinity transmitter, is an eavesdropping device used for planting within the telephone instrument itself. This device consists of a tone controlled switch, coupled with an audio amplifier and a microphone. Although it uses the telephone system, the device functions as a room eavesdropper as opposed to a telephone conversation interceptor. The infinity transmitter uses the existing telephone lines for conveying the surreptitiously acquired conversation. It is activated from an infinite distance by a tone generator similar to those used by answering machines. The tone is said to be a note produced by a harmonica, hence its alternative name.

In order to operate the infinity transmitter, the operative dials the target's telephone number, which can be local, interstate, or international. After dialing the number, but before the telephone rings, a tone is sounded into the operative's telephone mouthpiece. On the target end of the telephone, the infinity transmitter receives the audio tone and switches the device to answer this telephone electrically rather than physically. If this is performed correctly, the target telephone should not ring. This, in effect, means that the telephone is working even though the hand piece still remains on the cradle. Once operational, the surveillant may monitor the room conversation. If the subject attempts to use the telephone, the operative simply hangs up, and the device is electrically disconnected, returning the instrument to normal operation.

There are several other techniques directed at modifying the telephone instrument for eavesdropping. This form of electronic surveillance exploits the normal operation of the entire telephone system. By shorting or bypassing the hook switch on the instrument, the telephone becomes a live microphone. This technique is known as telephone "compromising."

Another dimension to electronic surveillance is systems that operate by using directional beams of light energy. These systems are based on the use of laser beams of either visible or infrared energy to convey their intercepted audio. They are reported to be quite reliable and virtually undetectable. The system consists of a laser light source, which focuses its beam on a window in a targeted room, and an optical receiving/ decoding device. The system operates by detecting minute vibrations of the reflected beam caused by the room's audio. Decoding of these vibrations reproduces the conversation or sounds within.

Interception of computer data can be accomplished in similar ways to eavesdropping on room conversations. With data interception, the legitimate user's data transfer is intercepted through a suitable wiretap or bug and either recorded or transmitted to a listening post as is done in the case of audio. With the interception of computer data, an agent monitors the data exchange from terminal to main computer, thereby tabulating input data and computer responses.

Pen Registers

There are also devices that allow operatives to collect data relating to the telephone number dialed by a target (known as a *pen register*) or to collect the telephone number that dials the target (known as a *trap-and-trace* device). The possession and use of these devices are governed by legislation, and laws vary from jurisdiction to jurisdiction.

Pen registers are usually manufactured to only record details such as the target's telephone area code, number, and, where an extension telephone is used, the number of the extension dialed. In the main, the reason why pen registers are not capable of collecting what is termed "transactional data" is that the data input by touchtone telephones—for instance an account number or a PIN number—would require another form of court order (i.e., warrant). Therefore, a pen register that is only capable of recording numbers avoids inadvertent breaches of an electronic interception warrant.

Internet-Based Intercepts

Apart from some university and government research libraries, the Internet is arguably the wealthiest source of information for analysts. It is not only a source for published material but also the private messages transmitted over the Internet (e.g., email). Businesses, governments, and individuals all create, access, or modify records of numerous descriptions that are located on computer servers connected to the Internet. Because these data exist in digital form on magnetic media, this realm is known as *cyberspace*.

In general, analysts can obtain data without a warrant if the data are placed in the public domain. This is analogous to the physical world where access is not restricted—public libraries, commercially published books and magazines, and newspapers. However, where access could be considered "private" by a court, then some form of order is likely to be needed. By and large, the laws pertaining to telephonic intercepts apply to intercepting data in cyberspace, and in many jurisdictions, laws have been enacted to cover online searches.

AUDIO SURVEILLANCE LIMITATIONS

For obvious reasons, it is impossible to determine the extent of electronic eavesdropping that is carried out by the intelligence community each year; although, from media reports, it appears to be quite widely used and not limited to any one intelligence type. If a target suspects he or she is the subject of a wiretap or bug, an audio countermeasure sweep is the usual way the target will try to determine if an operation is underway.

The sweep, however, will reveal devices operating at that given time only. It must be stressed that no room can be guaranteed to be proof against audio surveillance. In the past, even the most sensitive rooms in the United States embassy in Moscow have been reported to have been penetrated. Conducting sweeps at nonconstant intervals is the target's most effective way of countering an intelligence agency's audio surveillance. It is the most reliable way to check for, and clear, audio surveillance devices.

There are limitations, however, to this type of countermeasure. Firstly, with regard to telephones, even if the target's telephone instrument(s) appears to be clear at the time of a sweep, there is no way of determining whether the telephone of another party is under surveillance by inspecting the target's end of the line. There is also no technology available to date which can check for listening devices at, or beyond, the central telephone exchange. Secondly, there are some state-of-the-art devices and techniques used by intelligence agencies that may be undetectable because of their high level of technical sophistication.

Audio countermeasure sweeps are conducted by both specialist business counterintelligence firms and by private investigators. Such services are usually listed in the Yellow Pages of the telephone directory. A professional sweep usually includes both a thorough physical search—inspecting literally every inch of and every object in the suspected area—and an electronic sweep. The electronic sweep may utilize a broadband receiver (like those used by ham radio operators) and/or a specially designed field-strength meter to test for transmitters. Metal detectors can be used to hunt for bugs in nonmetallic objects and deeply planted devices in walls, floors, and ceilings. There are also a wide range of diagnostic meters used to test the telephone line voltage for the presence of wire taps.

PLACEMENT OF SURVEILLANCE DEVICES

If audio surveillance devices were to be placed in the target's property, it could be done by one or more of the following time-proven espionage methods:

- Friendly access;
- Surreptitious entry;
- Infiltration; or
- Secreted in a gift.

Friendly Access

Access to a target's building may be limited to employees and visitors who are known or have appointments. All other visitors may be screened and their identities verified prior to entry. People making deliveries, including mail deliveries and maintenance workers, may be handled in the same manner. Access to offices could be on a restricted, need-to-be-there basis. If visitor/staff traffic is heavy, a system of custom designed identity cards to be worn by employees would be an efficient method of establishing "friend" or "foe." Toilets and other isolated places might be checked at the end of the day's business for operatives hiding in the building.

Surreptitious Entry

Break-ins and burglaries are not an uncommon occurrence for businesses and private homes to experience. However, in June 1972, Watergate underscored the reality that break-ins are not only a method for acquiring cash and valuable physical assets but also are also a technique for information gathering.[6] In intelligence work, this technique is referred to as a *black bag operation*. Surreptitious entries are used to plant surveillance devices or to carry out other covert intelligence gathering activities. From the viewpoint of the operative, this is encouraging as short of creating a mini-fortress, there is nothing that will make an office 100 percent burglar-proof—even Buckingham Palace has had its intruder.

Infiltration Using a Pretext, Ruse, or Disguise

A *pretext* offers an operative acting on behalf of the intelligence analyst a plausible, common sense technique for obtaining confidential information. A pretext is any act of deception—ruse, subterfuge, ploy, trick, or disguise—that allows an agent to solicit information by a false reason. This includes entering premises for obtaining information or being in a place that an operative wouldn't otherwise have access to or permission for.

> "The art of using pretexts is a science and should be approached as one."[7]

Pretext should not be confused with the term *social engineering*, which has gained popularity in recent years. Social engineering is a slang term that commonly refers to an individual act of manipulation (usually for fraudulent purposes) to gain access to IT systems. This is vastly different from its true meaning, which is large-scale societal planning. The use of the term *social engineering* in this context is incorrect. The technique is nothing more than a ruse, subterfuge, or pretext. In fact, *pretext* is the term most used by private investigators—PIs rely heavily on this technique as a means of gaining information about their targets.[8, 9]

By Telephone

This method is used by operatives, usually on a one-time basis in order to obtain general information about a business. It is the safest and most innocuous type of infiltration to perpetrate. This type of infiltration is carried out by simply telephoning the target, then using a pretext, attempting to extract as much information as possible. Several calls could be made over a period of time.

On the surface, individual calls would appear to be unrelated, but each is designed to obtain specific pieces of information. Depending on the pretext and the number of pretext calls made, the depth of information an operative could gather might be limited to general information. The target might have acute security awareness, especially about unknown persons. If the target is suspicious, he or she may try to identify a telephone caller by requesting the caller's telephone number, then verifying it by using an online telephone directory before calling the operative back (known as *confirmation by callback*).

By Mail and Email

This is another low grade form of infiltration. Again, using a pretext, the operative will write to a target requesting information. Security-aware targets will look for the warning signs of a mail infiltration, such as the use of post office boxes, business name "fronts," and out-of-state addresses. As for email, free web-based accounts can raise the target's suspicions because these are non-verifiable accounts.

In Person

Direct personal infiltration of the target may follow pretext contacts by telephone and mail/email infiltration and physical surveillance. In this way, an operative can gather enough information to establish a credible cover for a direct penetration, or acquaint them with the information needed to recruit an agent (i.e., a proxy) to carry out the task.

Indirectly

This infiltration technique is complex to organize and run but can yield impressive results. Basically, an operative creates a covert business or organization designed to draw the target, the target's business, or a member of the target's staff. The bogus business is controlled by the operative. These covert enterprises can be as simple as a trade newsletter or as elaborate as a fully operational business.

Once established, the operative uses this cover to gather the information identified in the analyst's information collection plan. An example of this is the advertising of positions in a new and very attractive-sounding business. The business may offer a salary and fringe benefits package in excess of those offered in the market in order to entice the target. Once the target's curriculum vitae is received, it is analyzed for the desired information. If it does not disclose the information sought, additional information will be requested from the applicant and/or a personal interview conducted. The operative, or someone from this cover organization, would then "pump" the target for information.

Gifts

If a listening device is concealed in a gift that will be presented to the target (a form of a Trojan horse), it needs to be done expertly. There have been numerous cases where gifts have housed listening devices; the most notable was the 1952 presentation of the Great Seal of the United States to the American embassy in Moscow by the Russian government. A target that is security aware is likely to examine gifts well.

POSTAL MAIL COVER

A very useful data collection method is the postal mail cover. This investigative technique has been used by law enforcement agencies for many decades. It collects information by recording what is printed on the outer covering of an envelope or package via some form of photographic technique—photocopying, digital scanning, or digital photography. For the analyst, it is important to note that a mail cover operation does not involve reading or recording the contents of the postal item (e.g., letter or card); only the data on the outside are recorded.

Nonetheless, this simple and effective method of data collection can reap a wealth of information—return address, postmark related information (date and place of posting), and any description of the contents (e.g., "do not bend–photos enclosed," card, etc.) Because the mail is not opened, no search is being conducted, only a form of physical surveillance. The targets will never know that their mail is being monitored, so there is usually no administrative disclosure requirement by the agency that it is conducting surveillance.

Postal mail covers are usually bound by the laws or regulations of a nation's postal service, and these governance arrangements dictate how and when this technique can be used as well as how long the collection operation is allowed to proceed.

WASTE RECOVERY

This is a long-standing law enforcement and private investigator technique that is popularly referred to as *dumpster diving*.[10, 11] Despite the initial aversion to the thought of rummaging around in someone's waste material, this is a potentially rich and valuable source of information. Confidential material of all types can be found in waste—manuals, notes, letters, memos, reports, files, photographs, passwords, identity cards, receipts, schedules, itineraries, telephone numbers, and much more (including computer hard disk drives, USB flash drives, and a variety of data that have been backed up onto CDs/DVDs). The reason for this is that most people believe that once a piece of paper (or an old computer drive) is placed in a waste bin, it has "disappeared." They believe that no one would bother "getting dirty" searching through someone else's garbage.

Recovery can take place at any point between where the waste leaves the target's premises to, and including, the landfill site. If the recovery is to take place on the target's premises, be conscious that there may be legal issues associated with the operation, as a court could find that the material recovered was still in the possession of the target, and hence, a warrant of some kind was required.[12]

Information obtained via waste recovery was at one time considered high-value/low-cost because it yielded more benefit than what it cost to gather it. However, with its popularization in the press and cinema, waste recovery has become more difficult. Government agencies, businesses, and individuals regularly use document shredders and are more conscious of how and what they dispose. Security surrounding waste has improved with commercial-scale confidential document destruction becoming a service that is widely available.

KEY WORDS AND PHRASES

The key words and phrases associated with this chapter are:

- Aerial photography;
- Audio surveillance;
- Black bag operation;
- Bugs;
- Clandestine sources;
- Confirmation by callback;
- Covert photography;
- Covert sources;
- Dumpster diving;
- Electronic surveillance;
- Fixed surveillance;
- Friendly access;
- Infiltration;
- Informants;
- Internet-based intercepts;
- Moving surveillance;
- Open-source information;
- Operative;
- Optical surveillance;
- Paper trail;
- Pen registers;
- Physical surveillance;
- Postal mail cover;
- Pretext;
- Radio transmitters;
- Semi-open sources;
- Stakeout;
- Surreptitious entry;
- Surveillant;
- Tail;
- Telephone intercepts;
- Undercover agents;

- Waste recovery; and
- Wiretaps.

STUDY QUESTIONS

1. Explain the difference between obtaining information by covert methods as opposed to clandestine methods.
2. Explain the different types of covert data collection available to the analyst.
3. Describe some of the inherent limitations of covert data and the security precautions a target might employ to guard against penetration.
4. What are the advantages of covert data, and how could an analyst use these in practice? Give examples.

LEARNING ACTIVITY

Select a public building in your jurisdiction. Use an Internet-based aerial photography facility to create an electronic slide briefing for a notional surveillance team. Provide in your briefing details on access to and from the building, the surrounding terrain, and potential infiltration and exfiltration points. Note any limitations the maps may have as a way of better understanding what can be done with these types of sources.

NOTES

1. Richard Best and Alfred Cummings, *Open Source Intelligence Issues for Congress* (Washington, DC: Congressional Research Service, 2007).
2. J. Kirk Barefoot, *Undercover Investigation* (Springfield, IL: Charles C Thomas, 1975), 4.
3. T. J. Waters, *Class 11: Inside the CIA's First Post-9/11 Spy Class* (New York: Dutton, 2006), 118–19.
4. Graham Greene, *Our Man in Havana* (London: Heinemann, 1958).
5. Raymond Siljander, *Fundamentals of Physical Surveillance: A Guide for Uniformed and Plainclothes Personnel* (Springfield, IL: Charles C Thomas, 1977), 5.
6. G. Gordon Liddy, *Will: The Autobiography of G. Gordon Liddy* (London: Severn House, 1980).
7. Greg Hauser, *Pretext Manual* (Austin, TX: Thomas Investigative, 1994), 5.
8. M. Harry, *The Muckraker's Manual: How to Do Your Own Investigative Reporting* (Mason, MI: Loompanics Unlimited, 1980), 73–78.
9. Hauser, *Pretext Manual*.

10. John Hoffman, *The Art and Science of Dumpster Diving* (Boulder, CO: Paladin Press, 1993).

11. John Hoffman, *Dumpster Diving: The Advanced Course: How to Turn Other People's Trash into Money, Publicity, and Power* (Boulder, CO: Paladin Press, 2002).

12. Rick Sarre and Tim Prenzler, *The Law of Private Security in Australia, 2nd edition* (Pyrmont, Australia: Thomson Lawbook, 2009).

7

⚙

Data Collation Techniques

This chapter examines some of the most accepted techniques for collating information:

1. Brainstorming;
2. Nominal group technique;
3. Mind mapping;
4. Affinity diagrams;
5. Check sheets; and
6. Matrix diagrams.

INTRODUCTION

Filing information is simply cataloging data and storing it for retrieval. Although collation involves these two processes, it also entails some rudimentary forms of analysis. If the analogy of a pole-vaulter is used, the catapulting athlete can be seen as analysis and the run-up to the vault as collation. Analysis cannot be done without collation. And, as with the athlete's run-up to the high bar, how the run-up is executed will see the vault succeed or fail. If collation is done properly, then analysis will be enhanced.

In the main, collation aids the analyst to identify information that should be collected while highlighting information that was sought and hasn't been collected. And, it can show relationships that hitherto were unknown—especially if the collation technique is visual. As collation techniques are numerous, this chapter will discuss the most widely used.

BRAINSTORMING

Brainstorming has enjoyed widespread use among researchers for decades.[1] Although it has some limitations, it is arguably one of the most popular techniques for collating unstructured data. It can be used as a way to organize data post-collecting (e.g., when field operatives have seized large amounts of data from the target), or it can be used in the pre-collection phase to set up a structure for collecting (not to be confused with the information collection plan). Although brainstorming can be used in group settings (e.g., all members of the analytic section), it can be used in a solitary setting by individual analysts.

Brainstorming is a creative process that results in participants coming up with ideas that are a bit unusual. Nevertheless, it is this feature that allows the generation of numerous ideas in a lateral thinking process. Criticism is to be avoided, and evaluation of the pool of ideas only takes place at the end. Brainstorming sessions need only be short—5 to 20 minutes. It is important that the session facilitator does not let criticism creep into the discussion phase and keeps discussion focused on the defined problem (best posted in clear view in front of the group). Encouraging participants to have fun helps generate a wider range of ideas, but balance needs to be struck so that no one vein of thought is perused too far or for too long.

The ideas are recorded so that all participants can see the list, and this helps generate other ideas. If performed as an individual, a mind map can be used instead. If conducted as a group, an affinity diagram can help structure the ideas (see the Mind Mapping and Affinity Diagram sections, later in this chapter).

Some limitation occur when used by groups—participants can become distracted or disengaged mentally because they would rather be doing the work piling up on their own desks back in the office. In addition, some participants may inhibit open and creative discussion if there are personality clashes within the group. Yet still, some may feel self-conscious and may self-censor because they feel they are unable to contribute or that their contribution isn't as worthy as others made by the group. However, the use of brainstorming in groups can have a positive effect that lies outside the production of ideas: that is, it offers a platform for what is termed *buy-in* or *ownership* by those participating.

The 6-3-5 Method

There are several ways to stimulate creative thinking and the 6-3-5 method is one.[2] Using this method, the analyst convenes a group of six suitable people who are able to consider the brainstorming issue, or several groups of six can be convened simultaneously using this method. Each person is asked to generate three ideas within five minutes, hence the name 6-3-5.

Once this is done, participants pass their sheets of paper to the next person in the group. Each person views the other person's ideas, considers them, and then records three more ideas (within five minutes), thus building on the first three. The process is repeated until each person has had an opportunity to see all sheets of paper.

Variations can be made to this method if time or the number of contributors is constrained, say, a 3-3-2 method, or another variation.

Devil's Advocate

This can be a useful technique if used correctly during brainstorming. But it can also be a persona that could hold back creativity if used aggressively; its use is a matter of balance and good judgment. Simply, one member of the group acts as the devil's advocate by putting forward opposing ideas, options, or possibilities. It is useful because it steers thinking away from what could be a narrowly focused discussion, thus preventing "groupthink." Groupthink is where members of the group suppress alternative views because they think others may disagree or are embarrassed to voice dissent. Having a devil's advocate in the group can avoid the TINA (there is no alternative) syndrome by encouraging members to think wider.

NOMINAL GROUP TECHNIQUE

Nominal group techniques can be used as an alternative to brainstorming, especially where the participants may feel some constraint when it comes to articulating their thoughts in front of a group. The process is ideally suited as a means of circumventing a person(s) who may stifle discussion because it encourages passive participants.[3] Step by step:

1. Make the team of participants comfortable around a table, and supply them with writing materials;
2. Put the question under investigation to the group. It should be phrased as an open-ended question in order to encourage thought, for example: What are some of the ways Country Q could encourage its security forces to do [insert action]?;

Table 7.1. Ranking Ideas Method

	Dave	Betty	Kait	Lyndsey	Chris		Total
Issue A	1	5	4	1	2	=	13
Issue B	3	4	5	3	5	=	20
Issue C	3	5	3	4	3	=	18
Issue D	4	2	1	4	1	=	12
Issue E	1	5	2	1	2	=	11

3. Request that each participant consider the question for a minute or two and then individually brainstorm several possible ideas. The brainstorming is to be done silently with the participants' noting their ideas on a sheet of paper;
4. Call in the sheets of paper containing the participants' ideas. On a flipchart (or via a computer data projector), display the ideas so that all can see. During this process, eliminate duplicates by combining like with like. As with brainstorming, no criticism is allowed—all ideas are to be used. The facilitator can clarify ideas but does so by requesting guidance from the group before making any changes;
5. Request that all participants evaluate the ideas by individually (and anonymously) voting. This can be done several ways, but two commonly used methods are:
   - Ranking the ideas from most favored at the top to the least favored at the bottom. Then the rankings are combined and a total calculated for each idea. The ideas that score the most are the ones that deserve attention. Using a scale ranging from 1 to 5 (with 1 being the lowest and 5 being the highest priorities), table 7.1 illustrates this. It shows that Issue B would have the highest priority ($n = 20$), with Issue E the lowest ($n = 11$).
   - The other way is to use a weighting system. For instance, each participant can award 100 points across the ideas in any proportion. This is shown in table 7.2—Issue B has the highest priority ($n = 165$) in this example, with Issue E the lowest ($n = 25$).

Table 7.2. Weighting Ideas Method

	Dave	Betty	Kait	Lyndsey	Chris		Total
Issue A		25	40			=	65
Issue B	30	20	50		65	=	165
Issue C	30	30	10	50	35	=	155
Issue D	40			50		=	90
Issue E		25				=	25

MIND MAPPING

A mind map is a diagram representing a set of related ideas. Mind maps can take a variety of forms, but the two most popular are *hierarchical lists* and *spider diagrams*. In practice, it comes down to personal preference of the analyst which to use—some researchers think in terms of lists, while others are visual.

Table 7.3 shows an example of part of a mind map for this book, which was created by the author. It is evident that such a list lends itself ideally to became the book's table of contents (yes, a table of contents is a form of mind map). If an analyst prefers using a spider diagram, it can be converted into a list by grouping (and/or regrouping) the major ideas of the diagram into a hierarchical list. Likewise, a list can be converted into a spider diagram (see fig. 7.1).

"A mind map is the ultimate organizational thinking tool. And it is so simple! . . . Computers can be helpful when you mind map! Although it is still your brain that comes up with all the ideas, the latest software can allow you to draw a mind map on your screen. The advantages of this are obvious. You can save your mind maps in a file and then transmit this information to others. Computer mind maps allow you to store vast amounts of data in mind map form, to cross-reference that data, to shift branches around from one part of the mind map to another, [and] to rearrange entire mind maps in light of new information . . ."[4]

AFFINITY DIAGRAMS

Creating an affinity diagram is a method that analysts can use to cluster data items (ideas or concepts) along similar lines—hence, its name *affinity diagram* or *data cluster* (e.g., similar characteristics, qualities, attributes, parts, features, and so on). This clustering should not be confused with *classification*. In classification, items are assigned according to classes that have been predefined (e.g., in biology or agronomy). In clustering, the objective is to

Table 7.3. A Hierarchical List of Topics

The Fundamentals of Intelligence	*The Intelligence Research Process*
Intelligence versus information	Problem formulation
Intelligence defined	Literature review
Intelligence as knowledge	Methodology
Intelligence's consumers	Intelligence collection plan

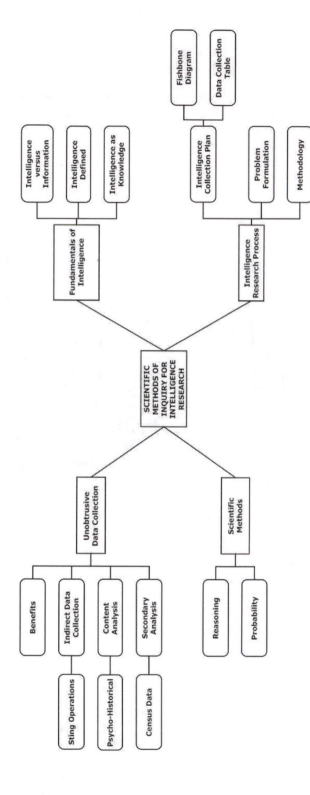

Figure 7.1. Spider diagram of topics.

take the data and break it down into smaller clusters, which can be better managed and, therefore, better analyzed.

Data clustering can be the follow-on step after a brainstorming session, or it can be used to help sort data that have been acquired by field operatives (e.g., from a search-and-seizure operation or material that has been acquired by an undercover agent or informant). Figure 7.2 shows an example of an affinity diagram of ideas that were generated by an individual who was brainstorming transnational organized crime issues. Step by step:

1. Separate the larger concept/issue/idea/problem into a number of smaller categories;
2. Display these categories via a convenient medium such as a board (e.g., whiteboard, pen-board, or marker-board), large flipcharts, or computer data projector;
3. Group (and/or regroup) the individual pieces of information (*factors*) into common clusters (as is done in mind mapping); and

Organised Crime Groups
- Definitions
- Theories
- History
- Geographic Differences
- Structures
- Functions

Financial Crimes
- Fraud
- Money Laundering
- Counterfeiting
- Securities Swindles
- Stock Manipulation

Smuggling Networks
- Drugs
- Guns
- Illegal Aliens
- Nuclear Material
- Flora and Fauna

Global Impact
- Political
- Social
- Economic
- International Security

Financial Sources
- Trafficking
- Bribery
- Police and Political Corruption
- Motor Vehicle Theft
- Murder
- Extortion
- Prostitution
- Gambling

Control Measures
- Global Cooperation
- Intelligence Sharing
- Cross-jurisdictional Operations
- Extra-legal Operations

Figure 7.2. **Affinity diagram of transnational organized crime issues.**

4. Give a heading to each of the clusters (this is done at the end of the process because if assigned initially, the name may change several times as more factors/data items are added, or it may give rise for a newly created cluster when, say, it is discovered that a cluster is too narrow to accommodate all the associated factors).

CHECK SHEET

Check sheets aid analysts to systematically compile and record data. These data can be either from historical sources or from observations as they occur, and the sheets can be designed and pressed into use very quickly.

The main advantage of using a check sheet system is that as data are entered into the system, a picture is built, making it increasingly clear what the facts are, as opposed to what opinions people are expressing about the problem, issue, or situation.

To create a check list, the analysts define the events or conditions that need to be observed. They then allocate who will collect the data and specify over what period of time (if it is a phenomenon that is occurring and being observed live) or from what data (e.g., historical records or files). Finally, they then design a check list sheet (which can be computerized using suitable software) and begin collecting data.

To illustrate this, take the example of a local village that reports seeing an increase in suspicious vehicle traffic passing to its south. It is rumored to be associated with a possible build-up of insurgents in the area. Field operatives are tasked to covertly record this traffic over a 24-hour period using a check sheet. Table 7.4 is an example of such an instrument.

Table 7.4. Example of a Check Sheet

Observed	Frequency
Passenger vehicles	涆 涆 |||
Light trucks	涆 ||
Tractors	涆 涆
10-wheeled trucks	||||
18-wheeled trucks	||

MATRIX DIAGRAM

Analysts can use a matrix diagram to show a many-to-many relationship between two data sets. This type of analysis is known as network analysis. In practice, it is a simple but effective technique, as demonstrated in detail in table 10.8 and figure 10.4 of chapter 10 ("Advanced Analytic Techniques"). You will note in these two illustrations that the various relationships are recorded in the matrix, but analysts can convert these data into a network chart (see fig. 10.4). Where a relationship exists between the data elements in the matrix, a symbol is inserted. A step-by-step description of how to perform this technique is provided in chapter 10.

KEY WORDS AND PHRASES

The key words and phrases associated with this chapter are:

- 6-3-5 method;
- Affinity diagrams;
- Brainstorming;
- Buy-in;
- Check sheets;
- Data clusters;
- Devil's advocate;
- Hierarchical lists;
- Many-to-many relationship;
- Matrix diagrams;
- Mind mapping;
- Nominal group technique; and
- Spider diagrams.

STUDY QUESTIONS

1. Explain the purpose of collation.
2. Describe some of the techniques available to the analyst for collating data.
3. Compare brainstorming with the nominal group technique by giving an example of how an analyst could use each.
4. Could brainstorming and the nominal group technique be substituted for one another? If so, could it be done in all circumstances or only in particular cases? Explain.

LEARNING ACTIVITY

Suppose an analyst is presented with the task of researching outlets for counterfeit name-brand goods. Use a mind map to collate information pertaining to the types of outlets that might exist in your jurisdiction and the location of same.

NOTES

1. Alex F. Osborn, *Your Creative Power: How to Use Imagination* (New York: Charles Scribner's Sons, 1948).

2. Helmut Schlicksupp, *Kreative Ideenfindung in der Unternehmung: Methoden und Modelle* (Berlin: de Gruyter, 1977).

3. A. L. Delbecq and A. H. Van de Ven, "A Group Process Model for Problem Identification and Program Planning," *Journal of Applied Behavioral Science* VII (July–August 1971): 466–91.

4. Tony Buzan, *How to Mind Map* (London: Thorsons, 2002), 4, 69.

8

⟨⟩

Basic Statistical Analyses

This topic presents the essential statistical techniques used in intelligence analysis:

1. Levels of data measurement;
2. Univariate analysis;
3. Calculating percent increases and decreases;
4. Per capita calculations;
5. Index numbers;
6. Ratios;
7. Rounding numbers;
8. Bivariate analysis;
9. Statistical significance; and
10. Problem solving using algebraic equations.

LEVELS OF DATA MEASUREMENT

In intelligence research, if variables are observed, the data gathered by these observations are organized according to numbers attributed to them by analysts. Assigning numbers to represent variables is done as a way of preparing these data for statistical testing. The numbers assigned to the data are the factors that determine the *level of measurement* and, hence, the kinds of statistical tests that can be applied.

There are four levels of measurement: nominal data, ordinal data, interval data, and ratio data. Each level of measurement represents an increase in the types of statistical tests that are permissible. As such, the level of measurement, it could be argued, is the foundational assumption for all statistical testing. That is, the data type determines which statistical test can be conducted. If this assumption is violated, it makes the results of any subsequent statistical test invalid.

For instance, if data were of a nominal scale, then only statistical tests designed to analyze these types of data can be used (say, for example, chi-square analysis). However, if the data were of a ratio scale, then any test at the ratio level and those for lower level data could be applied. This is because ratio data can be reduced to a lower level of measurement to be analyzed using an appropriate test. If the data are already at a lesser level of measurement, they cannot be converted to a higher level. Likewise, once the data are converted to a lower scale, unless the original data are retained in ratio form, they cannot be disaggregated.

Analysts also use what is termed *derived data*—but this is not the same as a level of measurement; it is information that is produced from combining or interrelating two or more data items to produce a new piece of information. For example, suppose an analyst has data on several events and also has data on when those events took place. Using a computer program, the analyst can produce new data based on an algorithm that requests "a list of all events that occurred before a certain date but after another specified date, as long as the event was not followed at any time by another type of event."

Nominal Data

Nominal data is the lowest level of measurement and comprises observations that can be placed in a group, for instance: Americans, Australians, Britons, Canadians, or New Zealanders (as such, it is sometimes referred to as a *categorical* scale). The chief attributes of nominal data are that there is no rank or order to the data—that is, one group is not "greater" or "more" than another group.

By itself, there is no "distance" between the groups as in higher level data. So, one cannot say that if you are an Australian, you are twice as "ethnic" as a New Zealander. All that can be said in this example is that one is in the Australian group, and the other is in the New Zealand group. Also, an observation can only be in one group, not multiple groups.

Ordinal Data

Ordinal data has the same attributes as nominal data, but in addition, it introduces the attribute of rank (sometimes referred to a *rank* scale). Rank in this sense suggests that the scale has some direction—say, from less to severe. Take, for instance, the crime of terrorism—three events could be ranked as to their severity: kidnapping, assassination, and bombing. One will note that they have the same attributes as nominal scale data in that observations can be placed in groups, but these groups have a relationship in how they are ranked. Because these observations can be ordered in this way, the level of measurement increases.

Nevertheless, even though there is direction to these data, there is no indication of distance between the data items. For instance, an analyst cannot say that assassination is twice as severe as kidnapping. However, the analysts can say that bombing is greater than kidnapping, that bombing is greater than assassination, and that assassination is greater than kidnapping.

Interval Data

Interval data has the same attributes as ordinal data, but it introduces the dimension of distance, or *interval*. That is, data measured using this scale will be able to demonstrate a common unit of measurement for all observations.

Take, for instance, the interval measure of time—a terrorist planted an improvised explosive device in a busy market place at 1:00 p.m., and it was detonated by remote control at 1:30 p.m. The analyst can conclude that there were two distinct events (nominal scale), one event followed the other (ordinal scale), and the time that elapsed between the two events was 30 minutes (interval scale). Because there is now a measureable distance, the interval scale allows the analyst to conduct arithmetic calculation on the data. One should note that there may be a zero point in this scale (e.g., 00:00 hours) but that zero is an arbitrary notion—it is not real.

Ratio Data

Ratio data is the highest level of data as it has all the attributes of interval data, but in addition, it has a real zero for the scale's starting point. Having a real zero starting point reference allows the analyst to calculate ratios between any two observations—data can be added, subtracted, multiplied, and divided.

By way of example, suppose the terrorist-improvised explosive device cited above was estimated to contain 15 kilograms of high explosive. Suppose also that another explosion that day was due to a device containing 30 kilograms. The analyst can conclude that there were two explosions (nominal scale), one used more explosives than the other (ordinal scale), the larger explosion was 15 kilograms greater (interval scale) as well as having twice the destructive power (ratio scale).

UNIVARIATE ANALYSIS

Univariate analyses are used when the analyst wants to simply describe a person, organization, location, or object that consists of a single dependent variable. Hence, univariate analysis is also known as *descriptive statistics*. Contrast this type of analysis with bivariate and multivariate analyses where two (bi-) or more (multi-) variables are analyzed (in such analyses, the variables may or may not be dependent upon each other). Univariate analysis can be used with the different levels of data as shown in table 8.1. The descriptive statistics that can be produced are listed in the right-hand column.

Frequencies

Constructing a frequency distribution is often an analyst's first task in analyzing data. This is done by counting the number of observations per category (nominal and ordinal level) or per score (interval and ratio level). The results can be described in the intelligence report's narrative as well as being displayed graphically. Figure 8.1 (ordinal level) and table 8.2 (ratio level) are examples of how terrorist event data can be presented. Figure 8.1 shows the number of terrorist events by organization before and after the 1986 U.S. air raid on Libya.[1] Whereas, table 8.2 shows the number of (notional) bombing events (X) terrorist groups were responsible for in Country Q.

Count

Count is the total number of values in a distribution (i.e., data set). Using the following distribution as an example, the count would be 11. These data could represent the number of people killed by roadside bombs:

4, 8, 10, 12, 15, 16, 19, 20, 24, 28, 31

Table 8.1. Examples of Descriptive Statistics

Measurement Level	Analytic Technique
Nominal	Frequencies
Ordinal	Frequencies, range, median, mode
Interval	Frequencies, range, minimum, maximum, median, mode, mean, weighted mean
Ratio	Frequencies, range, minimum, maximum, median, mode, mean, weighted mean

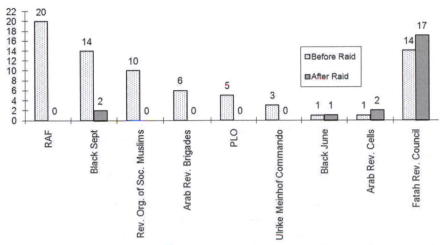

Figure 8.1. Terrorist events by Libyan-sponsored groups targeting Americans and U.S. property abroad.

Minimum and Maximum

The minimum is the lowest value in a distribution. The maximum is the largest value. For instance, using the same data set as above, the minimum number of people killed by roadside bombs and the maximum number were 4 and 31, respectively:

$$4, 8, 10, 12, 15, 16, 19, 20, 24, 28, 31$$

Range

The range is the difference between the minimum and maximum values with a distribution. This is calculated by subtracting the smaller value from the larger. Using the same example, the range would be 27:

$$31 - 4 = 27$$

Table 8.2. Frequency Distribution of (Notional) Bombing Events

X (score)	f (frequency)	fX
5	2	10
4	3	12
3	7	21
2	15	30
1	30	30
	N = 57	ΣfX = 103

Mean

The *mean*, also known as the arithmetic *average*, is used to average out quantities. It is calculated by adding all the values of a data set, then dividing the sum by the count of those numbers. Again, take for instance the distribution above: the total number of people killed was 187. Divide 187 by 11, and this equals 17:

$$(4 + 8 + 10 + 12 + 15 + 16 + 19 + 20 + 24 + 28 + 31) = 187 / 11 = 17$$

The mean can be used with both interval and ratio data; however, one of the disadvantages of using the mean is that it is affected by extremes at either end of the distribution. For example, if we use the notional roadside bomb data set and substitute 97 for the value 31, the resulting mean would be skewed to 23. Nevertheless, the mean is a useful statistical test as it can be used to conduct further tests—for instance, an analyst can compare the means of several different samples.

Weighted Mean

Another method used to average quantities is the *weighted mean*. It is used in cases where not all of the quantities are of equal importance. The formula for calculating a weighted mean is expressed in the following, where "bar-x" is the mean, w is the weight, x is the number, and Σ is the sum:

$$\bar{x}_w = \frac{w_1 x_1 + w_2 x_2 + w_3 x_3 + \ldots + w_n x_n}{w_1 + w_2 + w_3 + \ldots + w_n} = \frac{\Sigma\, w * x}{\Sigma\, w}$$

Suppose a situation where an intelligence analyst is trying to establish the average price paid for heroin on the street but on a national basis. Data from each state's capital city are compiled and are weighted according to the population of each city (i.e., using census data). This takes into account the fact that there are price variations at the same point in time between locations because illicit drug prices increase "as one moves away from the drug sources and prices are lower in larger markets."[2]

Suppose that there are three capital cities in Country Q. The price paid for heroin in one city is $148 per gram, another is $256, and the third is $300. The populations of the cities are 5 million people, 2 million people, and 1 million people, respectively. Therefore:

$$\bar{x}_w = \frac{(5)(148) + (2)(256) + (1)(300)}{5 + 2 + 1}$$

$$= \frac{1{,}552}{8}$$

$$= \$194 \text{ per gram}$$

If the analysts had averaged these data without using a weighting system to take into account the price variations caused by distances from the point of importation (the largest city with its port and international airport), they would have obtained $235. This is because the state with the smallest population that was far removed from the large port city—hence, the highest price—skewed the average upwards.

Weighted Grand Mean

There may be times when a *grand mean*, or the mean of the sum of all the means, is needed to be calculated. Such cases may arise where, for instance, a mean for heroin in each state's capital city has been calculated, but the national average is needed. To do this, the analyst totals the means for each city and then divides this figure by the number of cities. But because averaging the average can skew the results (sometimes considerably), a *weighted grand mean* is used. This is calculated by using the same formula for a weight average; however, instead of using x, the analyst uses *bar-x*, that is, the mean.

Median

The median is the middle value of a distribution. It is, therefore, the halfway point between those values that are greater than the median and the half that are less than the median. Using the following data set, the median is 16:

$$4, 8, 10, 12, 15, 16, 19, 20, 24, 28, 31$$

The median is less susceptible to extremes at the edges of the distribution. Again, take for example the notional roadside bomb data set we used previously, and again substitute 97 for the value 31, as was done for the mean, but now the resulting median would be 16, not the skewed 23 that was seen under the mean. Therefore, the median is generally more useful where there are extremes, and the mean most useful when the distribution is absent of such features.

Mode

The mode is the most frequently occurring value in a distribution. However, some data sets do not have a mode as is evident with the example distribution used so far—there is a single count for each number. If, by contrast, the following distribution was used, the mode would be 12:

$$4, 8, 10, 12, 12, 15, 16, 19, 20, 24, 28, 31$$

Sometimes data sets have several sets of numbers appearing in equal frequency. In such cases, these are referred to as being bimodal (two modes), trimodal (three modes), or in the case of more—multimodal. Bimodal:

$$4, 8, 10, 12, 12, 15, 16, 19, 20, 24, 28, 28, 31.$$

Trimodal:

$$4, 8, 10, 12, 12, 15, 16, 19, 19, 20, 24, 28, 28, 31$$

Of the three measures of central tendency—mean, median, and mode—the mode is the least useful unless the number of values represented by the mode form a large percentage of total distribution. Moreover, no further analysis can be conducted on the mode—an analyst cannot, for instance, compare modes of different samples as the result would make no sense at all.

CALCULATING PERCENT INCREASES AND DECREASES

Sometimes an analyst will need to express a number as a fraction of 100. This is usually done because the analyst needs to show how one quantity is related to another. The mathematical function used to do this is *percent* (meaning, per hundred). As an illustration, 54 percent is 54 / 100, or expressed as a decimal, 0.54. When a number increases, an analyst can calculate its percentage increase using the following formula:

percent increase = [(*new figure – original figure*) / *original figure*] × 100

When a quantity decreases, the analyst can calculate the percentage decrease by:

percent decrease = [(*original figure – new figure*) / *original figure*] × 100

It is important to note that both the formulas exhibit the following principle:

percent increase / *decrease* = (*change in figure* / *original figure*) × 100

This is because the analyst is calculating either the percent increase or decrease with the amount of change to the original figure as demonstrated in the following two examples:

1. The number of people attending political rallies in the city of Orrenabad rose from 16,000 to 22,000. What percent increase does this represent?

$$percent\ increase = (22{,}000 - 16{,}000) /$$
$$16{,}000 = 6{,}000\ /\ 16{,}000 = 0.375 \times 100 = 37.5\%$$

2. The number of political websites hostile to the new democratically elected government in Orren declined from 930 to 200. What percent decrease does this represent?

$$percent\ decrease = (200 - 930)\ /\ 930 = -730 /$$
$$930 = 0.785 \times 100 = -78.5\%$$

PER CAPITA CALCULATIONS

Per capita simply means per person. An analyst can calculate the per capita occurrence of x where x is the issue under investigation:

$$per\ capita = (total\ x\ /\ population) \times rate$$

So, using an example of suicides, the per capita figure would be the total number of self-inflected deaths divided by the population. In this case, the population could be either the population of the town or city, the region or state, or the country. This statistic can then be used as a means of comparison to other towns or cities or from country to country in a meaningful way.

One of the detractions of per capita calculations is where the figure is very small; in such cases, the ratio becomes meaningless. In such cases, the analyst can use per 100,000 in the population, 10,000, or even 1,000 as alternatives.

By way of example, take the notional regional town of Isobel. The town recorded 200 residential burglaries during the year, and it had 100,000 households. Therefore, the burglary rate would be 0.002. This is not a very meaningful figure. However, if it is calculated per 1,000 in the population, then it is 2 burglaries per 1,000 households.

$$(200\ /\ 100{,}000) \times 1{,}000 = 2$$

By using a rate per 1,000 instead of a per capita rate, in this example, the figure can more easily be understood.

INDEX NUMBERS

A benchmark can assist analysts when comparing changes in different kinds of data, and in statistical terms this is referred to as an *index*. "Since index numbers show percent changes rather than arithmetic change, the size of the data and the units of measurement are not important."[3]

As a demonstration, the rate of inflation can be compared with the production of goods by using the consumer price index with the industrial production index. "Although prices are measured in dollars and production is measured in physical volume (number of cars, tons of coal, etc.),"[4] the two can be easily compared using this method.

As part of a study into heroin trafficking in Australia, Prunckun[5] used index numbers to control for confounding variables. His study collected data relating to Australia's population, its rate of inflation, and its number of sworn police officers. These data were gathered as a means of comparing or adjusting the rate of change observed in variables. For instance, population data were used to adjust methadone and heroin usage rates relative to "the population in the age groups in which opiate users are found (namely 15 to 44 years)."[6] The inflation rate was helpful in adjusting the illicit drug prices, and the police staffing data were helpful in comparing changes in police numbers with that of some of the law enforcement workload data. The formula for calculating an index is:

$$index = (current\ value\ /\ base\ value) \times 100$$

Suppose that Country Q had 5 warships in 2005. By 2010, intelligence indicated that this number had grown to 20 (with the intervening years being 6 in 2006, 9 in 2007, 14 in 2008, and 17 in 2009). How has this number changed between 2005 and 2010?

The base period is 2005, and the *base value* is 5, which is given a value of 100 that corresponds to 100%. The index is then calculated for the period following the base period using the above formula and is displayed in table 8.3. The index number shows that there was an increase of 300% in warships since the base period (i.e., 400% − 100% = 300%).

RATIOS

An analyst who wants to compare two things can use a ratio. A ratio is an expression of something *to* something else, for instance: a ratio of small arms to soldiers. Ratios are expressed in several ways: either as a fraction (3/4) or using the word *to* (3 to 4) or using a colon (3:4). To illustrate this, say, an analyst observes that there are 30 soldiers in a combat unit of a foreign country. One is assigned to radio communications. Later, intelligence

Table 8.3.   Example of Index Numbers for Warships

Year	Number of Warships	Index Number
2005	5	100
2006	6	120
2007	9	180
2008	14	280
2009	17	340
2010	20	400

indicates that this unit has added a second radio operator, but the size has grown to 60 soldiers. Despite the changes in overall numbers, the ratio of radio operators to soldiers within the unit has remained the same—30:1 before and 60:2 after (which can be simplified by mathematical reduction to 30:1).

ROUNDING NUMBERS

Rounding numbers is a common occurrence where the decimal fraction makes no sense or adds confusion. Sometimes an estimate is all that is needed to demonstrate a point or make a prediction. In these cases, rounding takes place. Step by step:

1. Determine the unit that is required to be rounded off—thousands, hundreds, units, or a decimal fraction;
2. Examine the number to the right of the unit to be rounded off;
3. If this number is 5 or more, add an additional unit;
4. If the number is less than 5, subtract a unit.

For instance, if rounding 394.6 to the nearest hundred, it would be 400. If rounding 394.6 to the nearest tens, it would be 390. But, if rounding 394.6 to the nearest unit, it would be 395.

BIVARIATE ANALYSIS

Chi-square

Chi-square is one of the more useful statistical tests for intelligence analysts because it can be used on nominal level data. It has few assumptions and is, therefore, straightforward to apply and interpret.

Chi-square is a nonparametric test of statistical significance. The term *nonparametric* refers to statistics that deal with variables that are without assumptions as to their form or their distribution parameters. It returns a statistic that reflects the "goodness of fit" or the difference between the observed frequency and the expected observations according to a model hypothesis (H_0).

The power of the chi-square statistic is that it will tell the analyst whether the actual distribution occurred by chance or was likely to be the result of the interaction of the independent variable according to a level of confidence (e.g., .01 or .05). Later in this chapter, the importance of setting the appropriate level of confidence is discussed in more detail. There are only three requirements for using chi-square and these are:

1. Any level data can be used;
2. There must be more than five observations per category (if there are less than five observations, they will be meaningless); and
3. The observations must be independent.

Unlike some other statistical tests, there is no direction indicated by chi-square, so once the result is obtained, an inspection of the data is required to determine this. Consider the following example of a single sample chi-square: Analysts are concerned about whether the numbers of terrorists in a target organization's cells in North America are equal to those in the organization's Asia region. The observed frequencies are arranged in table form (referred to as a *contingency table*). If the data were represented by two independent samples (that is, in the form of a 2 × 2 contingency table), then the convention is to list the independent variable (x) along the top of the table with the dependent variable along the side (y).

Chi-square can also be used for cases involving two independent samples. For instance, suppose an analyst wanted to know if there were significant differences between insurgents who were young males as opposed to older males and whether these people had previously been involved in criminal activity before joining the insurgency. H_0: The number of terrorists in North American cells and Asian cells are equivalent. H_1: The number of terrorists in Asian cells is more. The observed distribution is:

North America	86
Asia	120
Total	206

The expected distribution under the H_0 is for 50 percent to appear in each:

North America	103
Asia	103
Total	206

The formula for calculation of chi-square is as follows:

$$X^2 = \Sigma \frac{([A_1 - E_1] - .5)^2}{E}$$

X^2 = chi-square
A = actual or observed frequency
E = expected frequency
$-.5$ = Yates' correction for continuity

The calculations follow; however, many software spreadsheet packages contain the chi-square function. It is a simple matter of entering the data into the spreadsheet, indicating the degrees of freedom, and activating the chi-square function from the menu options. The spreadsheet will return the chi-square statistic (here it was done manually and is shown in table 8.4).

Yates' correction for continuity is applied in instances where the degrees of freedom are equal to 1 or where the observed frequencies are less than 10. This prevents an overestimation of statistical significance. The *degrees of freedom* are calculated thus: $df = C - 1$ (where C represents categories). So, in this example it would be $2 - 1 = 1df$.

In chi-square analysis, the *critical value* is the threshold that determines at what point an analyst would not reject the null hypothesis. Using the critical values contained in the appendix, we observe that a value of 5.28 is greater than the required 3.84 at the .05 level ($1df$), so we can reject the null hypothesis—numbers of terrorists that exist in the Asia cell are greater (i.e., we accept H_1). This means that there would be less than 5 chances in 100 that a result like this would be obtained if random variation was the only explanation.

As can be deduced from the formula, the chi-square statistic is the product of each of the individual frequencies. Therefore, each frequency contributes to the final chi-square statistic. If a given frequency is greatly different from the expected frequency, then its contribution to the chi-square statistic can be expected to be large. Conversely, if the frequency

Table 8.4. Manual Calculations for Determining Chi-Square

	North America	Asia	Total
Actual/observed	86	120	206
Expected	103	103	206
$(A - E) - .5$	16.5	16.5	
$([A - E] - .5)^2$	272.25	272.25	
$([A - E] - .5)^2 / E$	2.64	2.64	
Σ	5.28		

closely aligns itself to the expected frequency, then the contribution of that frequency to chi-square will be small.

It is clear that large chi-square statistics indicate that the contingency table contains a frequency(s) that differs noticeably from the expected frequency(s), but the statistic will not be able to point to the frequency(s) responsible for the elevated chi-square. It can only indicate that such a frequency(s) is present. When an excessive result is obtained, it requires the analyst to examine the table in order to determine which frequency(s) is the cause.

In the body of the analyst's report, it should discuss the conclusions drawn from these results, what they mean, and the implications they have for the study. When discussing these issues, refer directly to the table(s) so the reader is not left in doubt about any aspect of the conclusions.

Scatter Plots

A scatter plot (sometimes termed *scattergram* or *scatter diagram*) is used to visualize relationships between two data sets (i.e., paired samples). In doing so, the two variables do not have to have a dependent and independent relationship as plotting will demonstrate the degree the variables are correlated. As such, plotting data using this method is synonymous with *regression analysis*.

Scatter plots are able to reveal a number of relationships in addition to the degree of correlation, for instance: a positive relationship, a negative

Figure 8.2.   Positive correlation.

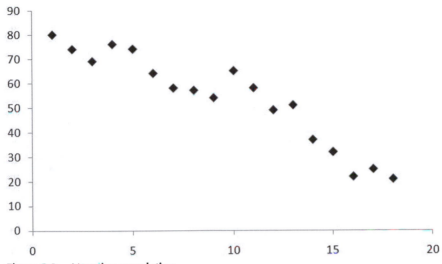

Figure 8.3.   Negative correlation.

relationship, and a null relationship. These are illustrated in the following series of figures. Figure 8.2 shows the pattern of dots sloping from the lower left to the upper right, thus demonstrating a positive relationship. Figure 8.3 shows a pattern of dots sloping from the upper left to lower right suggesting a negative correlation. The "tighter" the dots are in forming a straight line the more correlated the data are—see figure 8.4 for an

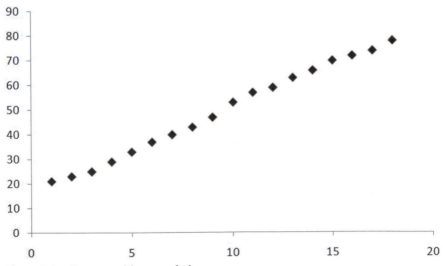

Figure 8.4.   Strong positive correlation.

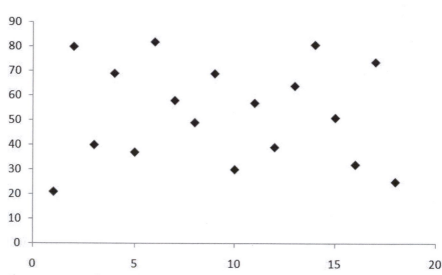

Figure 8.5. Nonlinear relationship.

example of a perfectly corrected paired sample. Finally, figure 8.5 shows a nonlinear, or null, relationship.

An analyst can quickly construct a scatter plot either by hand or by using a software package (usually a spreadsheet program). The first step is to draw a chart with a horizontal axis (x) and a vertical axis (y) and place appropriate and proportional numeric gradations along the axes. Data are then plotted on the graph. If doing this by hand where values are repeated, the analyst can indicate the repeated values by placing a circle around those points. If using a software package, the program will generate the chart once the data items are entered into the spreadsheet. Follow the menu options to produce the scatter plot.

The chief advantage of a scatter plot is that it will indicate the strength of the relationship between two variables. The stronger the relationship the more likely a change in one variable will result in a change in the other (but will not predict *cause*).

STATISTICAL SIGNIFICANCE

Type I and Type II Errors

The concept of statistical significance is important to the analyst as it makes clear what likelihood there is for an error in judgment. At the center of this process is the confidence level chosen by the analyst—this is referred to as the *alpha-level* or *p-value*, for example: .05, .01 (or another level either higher or lower). Regardless of the level chosen, there is the risk that the null hypothesis could be rejected when it should have been accepted or vice versa—

accepted when it should have been rejected. The risk of error increases as the confidence level gets smaller.

For instance, a *p*-value of .01 means that there would be less than one chance in one hundred that a result like this would be obtained if random variation was the only explanation (i.e., a 99% chance of being true). A *p*-value of .05 indicates that there is a 95% chance of being true. And, a *p*-value of .001 indicates there is 99.9% chance of being true.

Such errors are referred to as type I errors and type II errors. The former are false positives, and the latter are false negatives.

Suppose an analyst has selected a confidence level of .05; in this case, the analyst could more easily reject the null hypothesis and declare that there is a statistically significant difference in the data than what would have occurred if he or she had selected the .01 level. However, the analyst would be wrong in this conclusion 5 percent of the time. If, however, the analyst selected a confidence level of .01, he or she would be wrong only 1 percent of the time.

Type I errors are when some phenomenon is observed and the conclusion is drawn that there is a difference when in fact none exists—this is why they are viewed as false positives. A type II error is a false negative—accepting the null hypothesis when it should have been rejected. This is where the analyst fails to observe a difference and accepts the null hypothesis.

What is the impact on the analyst's findings of type I and type II errors? A type I error could be described as a "false alarm"—sending a signal that the observed phenomenon is worthy of note (and presumed action) when in fact it is a mistaken conclusion. The analogy could be a miner finding pyrite (i.e., fool's gold) instead of gold. In contrast, a type II error could be attributed to a lack of sensitivity of the data collection instrument or an omission of some sort so that the difference was not detected. If the miner analogy is used again, the miner would have thrown out the gold with the tailings from the mine.

From this discussion, the analyst can see that in any research project, they are at risk of making one of these two types of errors. So, which error is "worse" to make? It depends on the question that has been asked by the decision maker. Both can be equally devastating as can be seen in the following two examples:

- If the decision maker asked if there are any weapons of mass destruction in Iraq, then committing a type I error will have serious repercussions if the decision maker is planning a military invasion (i.e., wrongly concluding that such weapons existed when there were none).
- If, however, the analyst is asked whether the United States faces an attack by a terrorist group (e.g., in the manner of al-Qaeda), then a type II error needs to be avoided—overlooking a genuine threat to national security could result in a September 11, 2001-style attack.

PROBLEM SOLVING USING ALGEBRAIC EQUATIONS

Algebraic equations are used in a range of research activities to solve a myriad of problems. In essence, algebraic equations are formulae-based models that represent situations in the kinetic world. Rather than try to canvas the countless ways algebra can be used, we will draw on an indicative example to demonstrate the power of this type of analysis.

Consider a situation where an operational commander needs to be advised as to the estimate of the arrival time of enemy troops (in a military intelligence context) or the arrival of a motorcycle gang that is on the move (in a law enforcement intelligence context). Using algebra to solve this dilemma becomes an exercise in predictive intelligence, albeit one that is purely operational. To solve this problem the analyst:

- Notes the location of the troops/motorcycle gang at present;
- Notes the location of where the troops/motorcycle gang are expected to arrive;
- Calculates the distance between the two points;
- Notes the rate of advance (e.g., from information received from a reconnaissance unit, an airborne observer, or covert surveillance);
- Calculates the approximate arrival time for the troops/motorcycle gang, taking the distance between the two location points and dividing it by the rate of advance ($T = D / R$).

In this example, suppose the distance between the two points of interest is 200 kilometers. And, suppose that the persons of interest are advancing at a rate of 50 kilometers per hour. Therefore, 200 km / 50 km per hour = 4 hours. The intelligence analyst can then brief the commander that the estimated arrival time of the persons of interest will be in approximately 4 hours.[7]

This type of estimate can be applied to business intelligence and private intelligence too. In the case of the former, analysts could estimate the market arrival of a competitor's new model Gizmo. If the rate of production per unit is known (or can be estimated based on previous information or similar processes), then the timeframe for delivery can be calculated by inverting the equation used above and amending the variable labels to reflect the events:

delivery date = number of units needed $\times$ *rate of production*

In the case of private intelligence, an analyst can use algebraic equations to, say, deduce that it was feasible for the suspect in an insurance fraud matter to set the fire in question. Using the formula cited in the distance traveled example above, the analyst could use it to show that the person of interest was capable of traveling, without exceeding the speed limit, from his location to the site of the alleged arson and back within the times given by witnesses.

KEY WORDS AND PHRASES

The key words and phrases associated with this chapter are:

- Algebraic equations;
- Average;
- Base value;
- Bimodal;
- Bivariate;
- Categorical data;
- Chi-square analysis;
- Contingency table;
- Critical values;
- Descriptive statistics;
- Frequency;
- Index numbers;
- Interval data;
- Level of confidence;
- Levels of measurement;
- Maximum;
- Mean;
- Medium;
- Minimum;
- Mode;
- Multivariate;
- Nominal data;
- Nonparametric;
- *P*-value;
- Per capita;
- Percent decrease;
- Percent increase;
- Range;
- Rank data;
- Ratio data;
- Ratios;

- Regression analysis;
- Rounding;
- Scatter diagram;
- Scattergram;
- Scatter plot;
- Statistical significance;
- Trimodal;
- Type I error;
- Type II error;
- Univariate;
- Weighted grand mean;
- Weighted mean; and
- Yates' correction.

STUDY QUESTIONS

1. What are the four levels of measurement in statistics? Describe the attributes of each level.
2. List the different types of univariate analysis available to the intelligence analyst.
3. Describe the differences between mean and weighted mean. Give an example of how an analyst might use each.
4. What is the difference between a type I error and a type II error? Explain.

LEARNING ACTIVITIES

1. Suppose ocean-going freighters regularly use the straits off the coast of Country Q to save travel time. Suppose also that these straits are the subject of pirate attacks. You, as an intelligence analyst, have been asked to calculate the decrease in international shipping using these straits after a certain date. The daily number of ships before was 416 and after was 232. What was the percentage decrease due to hostile activity?
2. Community elders in Country Q are concerned about attacks on street markets—they suspect that they are not random and that this could signify growing factional tensions. The intelligence cell attached to the nation's paramilitary police has divided the country into five equal regions and collated the attack data accordingly. This information revealed the following distribution:

Region	Number of Attacks
1	26
2	19
3	31
4	17
5	11

Using a chi-square test of significance, determine whether it is likely that this distribution of attacks is by chance (random) or that other forces may be at play. Use a 0.05 confidence level. What can you conclude from these results?

NOTES

1. Henry Prunckun, "Operation El Dorado Canyon: A Military Solution to the Law Enforcement Problem of Terrorism—A Quantitative Analysis" (master's thesis, University of South Australia, 1994), 47.

2. Jonathan Caulkins, "What Price Data Tell Us about Drug Markets," *Journal of Drug Issues* 28, no. 3 (Summer 1998): 602.

3. Ester H. Highland and Roberta S. Rosenbaum, *Business Mathematics*, 3rd ed. (Englewood Cliffs, NJ: Prentice-Hall, 1985), 420.

4. Highland and Rosenbaum, *Business Mathematics*, 420.

5. Henry Prunckun, "Chasing the Dragon: A Quantitative Analysis of Australia's Law Enforcement Approach to Combating Heroin Trafficking—1988 to 1996" (PhD diss., University of South Australia, 2000).

6. Commonwealth Department of Human Services and Health, *Review of Methadone Treatment in Australia* (Canberra, Australia: Australian Government Publishing Service, 1995), 29.

7. U.S. Department of the Army, *FM 34-2-1: Tactics, Techniques and Procedures for Reconnaissance and Surveillance, and Intelligence Support to Counterreconnaissance* (Washington DC: Department of the Army, June 1991), 2–20.

9

❦

Presenting Statistical Results

The key concepts that will be discussed in the chapter are:

1. Graphs;
2. Tables; and
3. Stem-and-leaf plots.

GRAPHS

Graphs are used to display numeric data pictorially; in effect, they are symbolic representations. In practice, the term *graph* is used interchangeably with the term *chart*. Intelligence analysts use graphs to display information that would be too complex to do so in narrative form. Therefore, the graph offers the analysts a simple, concise format for conveying the gist of their results to the decision maker.

The most commonly used (and easily recognized by nonanalytic users) are the *pie chart*, *bar chart*, and *line chart*. A pie chart is a circular figure divided into segments that resemble a pie. Each "slice" of the pie represents a particular data item. The size of the overall pie is 100 percent, with each slice being a representation of a proportion of the total.

A bar chart uses vertical bars (but sometimes they are laid out horizontally) to show the relationship between data across categories. Bar charts are used where the data items are categorical (i.e., independent of each other), for instance: the number of rockets, automatic weapons, and hand

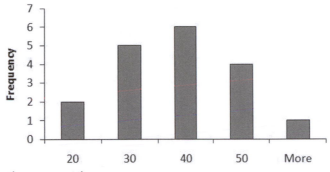

Figure 9.1.   Histogram.

grenades that a guerrilla force may have in its possession (a pie chart could be used for the same purpose as it also represents individual data items). The bars are not drawn contiguously, as in a histogram (see fig. 9.1), but with separations between them to demonstrate they are not related by time or score interval. If, however, the bars are joined in a contiguous manner, then it is referred to as a *histogram*. A histogram is a picture of a grouped distribution, which shows the shape of the distribution.

A *frequency polygon* serves the same purpose as a histogram, but it only shows the midpoint of the intervals of the scores. It can be displayed as either bars or dots. If shown as a series of dots, each dot is connected by a line (see fig. 9.2). In this regard, the line becomes a bar chart but with the bar image suppressed—they are merely represented by dots joined by a line.

Figure 9.2.   Frequency polygon.

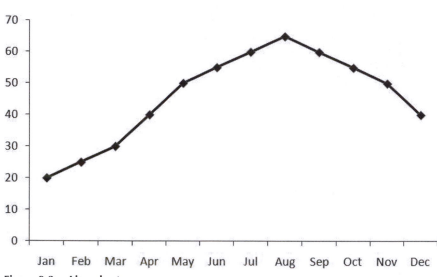

Figure 9.3. Line chart.

In contrast, a line chart is used where the data are required to be displayed showing passage of time (time series), say, the number of rocket propelled grenades an insurgent group has held in its cache month by month (see fig. 9.3).

Some forms of graphs use pictures or symbols in place of bars, columns, or lines as a way of reinforcing the message. These are termed *pictographs* and lend themselves to briefings where the analyst is presenting his or her results via a data projector on a large screen. However, before an analyst uses such a technique, thought should be given to whether its inclusion will trivialize the presentation. This is because the images that are used can be construed by decision makers as frivolous, and therefore, they lessen the impact of the message being conveyed.

SOME TIPS FOR CONSTRUCTING GRAPHS.

- Give graphs captions that reflect the dependent and independent variables being displayed.
- The X axis displays the independent variable, while the Y axis the dependent variable.
- Label both axes, and assign an appropriate unit of measurement.

TABLES

A *table* is a data set arranged in columns. When constructing a table, convention is to only use horizontal lines. Some software packages place vertical lines into tables, but strictly speaking, this is not correct. An example of a properly laid out table is shown in table 9.1.[1]

STEM-AND-LEAF PLOTS

As an alternative to displaying data in a histogram, the analyst can use a stem-and-leaf plot. Stem-and-leaf plots are constructed by creating a *stem* (left-hand column) that contains the tens-unit digits. The *leaves* of the plot are the integer-unit digits for each corresponding tens-unit and listed in the right-hand column. Once constructed, the horizontal leaves in a stem-and-leaf plot correspond to the vertical bars of a histogram.

Stem-and-leaf plots can be used for decimal numbers also. They are produced by using the integer portion of the data as the stem and the digit after the decimal point as the leaf. If the decimal portion has two figures, the analyst rounds the last digit so only one decimal place remains. The following shows a stem-and-leaf plot using whole numbers:

```
5: 3
5: 4 4
5: 5 5 5
5: 6 6 6 6 6
5: 7 7 7 7 7 7 7
5: 8 8 8 8 8 8 8 8 8
5: 9 9 9 9 9 9 9 9 9 9 9
6: 0 0 0 0 0 0 0 0 0
6: 1 1 1 1 1 1 1
6: 2 2 2 2 2
6: 3 3 3
6: 4 4
6: 5
```

Table 9.1. Worldwide Terrorist Events by Severity Level Pre- and Post-1986 U.S. Air Raid on Libya

Severity Levels	Terrorist Events		
	Before Raid	After Raid	% Change
High	541	458	−15.3%
Medium	170	155	−8.8%
Low	145	244	+68.3%
Totals	856	857	+00.1%

And the following is a stem-and-leaf plot using decimal numbers:

5: 3
6: 3 4
7: 4 5 6 7
8: 2 6 7 8 8
9: 2 3 4 5 5 6 8 9
10: 1 1 2 3 5 5 6 8 8 9
11: 0 1 2 3 3 4 5 5 6 7 7 9
12: 1 2 3 3 4 4 5 6 8
13: 1 1 2 3 5 9
14: 2 4 7 9
15: 4 3 7
16: 5 7
17: 1

Note that the length of the leaves shown in the leaf-and-stem plot for whole numbers corresponds to the frequencies in table 9.2 when represented in a frequency table. To construct a frequency table, the analyst arranges the data by categories, increments, or—as is the case with table 9.2—groups.

Table 9.2. **Example of a Frequency Table**

Groups	Frequency
51–55	6
56–60	44
61–65	18

TRAPS TO AVOID

Here are a few things to keep in mind when constructing graphs:

- Do not be tempted to enhance a graph by using what could be considered a gimmicky feature contained in some software packages—such as three-dimensional charts. Features like this trivialize your research and distract the decision maker's attention away from your study's results and focus too much on the eye-catching graphical image;
- Be mindful of inadvertently distorting chart results by setting the baseline value (i.e., the value at the bottom of the vertical, or *y*, axis) to a score other than zero; and
- Do not mistakenly display categorical data as a frequency polygon (i.e., along the horizontal, or *x*, axis)—instead, present these data as a bar chart.

KEY WORDS AND PHRASES

The key words and phrases associated with this chapter are:

- Bar chart;
- Chart;
- Frequency polygon;
- Frequency table;
- Graph;
- Histogram;
- Leaves;
- Line chart;
- Pictograph;
- Pie chart;
- Stem;
- Stem-and-leaf plot;
- Table;
- X axis; and
- Y axis.

STUDY QUESTIONS

1. Explain why an analyst would use graphs to display statistical results.
2. What are the types of visual representations available for intelligence analysts to use?
3. Discuss the difference between a histogram and a frequency polygon.
4. Describe the attributes of a correctly laid out table of statistical data.
5. When would an analyst use a stem-and-leaf plot? Give an example.

LEARNING ACTIVITY

Suppose an intelligence analyst is preparing a briefing to industry executives of the occurrences of trademark violations in a major city known for its high international tourist traffic. Over the past year, field operatives have noted the following numbers of individuals trading in counterfeit trademarked goods for each month starting in January and ending in December: 27, 35, 17, 13, 42, 37, 52, 48, 20, 24, 12, and 30. Select an appropriate method of displaying these data for an electronic slide presentation, and prepare the slide.

NOTE

1. Henry Prunckun, "Operation El Dorado Canyon: A Military Solution to the Law Enforcement Problem of Terrorism—A Quantitative Analysis" (master's thesis, University of South Australia, 1994), 52.

10

ॐ

Advanced
Analytic Techniques

A number of sophisticated analytic techniques will be discussed in this
chapter:

1. SWOT analysis;
2. PEST analysis;
3. Force field analysis;
4. Pareto analysis;
5. Analysis of competing hypotheses;
6. Fishbone analysis;
7. Morphological analysis;
8. Perception assessment analysis;
9. Timeline analysis;
10. Network analysis;
11. Telephone record analysis;
12. Event and commodity flow analysis;
13. Genealogical analysis; and
14. Financial analysis.

INTRODUCTION

One issue that is common to advanced statistical analyses is "customer acceptance." Due to the abstract nature of the mathematical formulae of techniques, like multivariate analysis, decision makers can feel troubled when
asked to trust assessments based on such methods. However, this chapter
describes a number of advanced analytic tools that can be used to make

sense of unstructured data in a range of intelligence settings that are easy to understand by decision makers and easy to employ by the analyst.

The reason for discussing a variety of tools is that each tool tends to be problem specific. That is, some tools work better with certain types of data or with certain types of issues, and others work better in other circumstances. Some tools are used for strategic analysis, while others are used for tactical or operational problems.

But just because an analyst has examined data using one of these tools does not mean that the results can be taken as absolute—analytic techniques are simply methods that allow analysts to form judgments (i.e., defensible conclusions) in a way that is transparent to the reader of their reports. Their analyses are able to be repeated in order to demonstrate the validly and reliability of the methods used (i.e., in keeping with the tenets of scientific research). The point is that analytic tools are not replacements for sound critical thinking or the application of professional judgment, but they aid both.

> There is little doubt that the analysis of qualitative data enjoys no consensus about what technique is used.

SWOT ANALYSIS

Arguably, an analysis of strengths, weaknesses, opportunities, and threats (SWOT) is one of the most popular analytic tools used by intelligence analysts. This is for two reasons: 1) it can be used with a variety of unstructured data (qualitative data from either primary or secondary sources), and 2) the focus of the research is not variable dependent—it can be either the target or the agency conducting the operation against the target. The technique was devised for long-range business planning, but it can be applied to a variety of issues in the intelligence realm that are either strategic or tactical. Or it can be used to analyze information in order to build a profile or help understand the current situation (e.g., situational analysis). A SWOT begins with the analyst's defining the *end-state*, as it is called in a strategic setting or *objective* if it is tactical.

Using any one of several idea generation or data collation methods, the analyst populates each of the four quadrants of the SWOT matrix with the data (although a matrix typically displays a SWOT, SWOT analyses can be laid out in any way that is suitable for the analyst). For strategy assessments, it is advantageous to have a broad view of the issue and, hence, employ a multidisciplinary team approach (i.e., an ideas workshop) to consider each of the four factors. A tactical assessment could be done by

Table 10.1. SWOT Analytical Matrix

	Analysis of Strengths, Weaknesses, Opportunities, and Threats	
	Supportive	*Detrimental*
Internal	*Strengths* are the attributes associated with the [issue/problem/agency/etc. under investigation] that are conducive to achieving the end-state.	*Weaknesses* are the attributes associated with the [issue/problem/agency/etc. under investigation] that are detrimental or may prevent achieving the end-state.
External	*Opportunities* are the conditions [legal/criminogenic/social/economic/political/psychological/etc.] that would assist achieving the end-state.	*Threats* are the conditions [legal/criminogenic/social/economic/political/ psychological/etc.] that might be detrimental to the way the agency carries out its operations.

an analysis based on the data collected in the lead-up to an operation or during an operation.

Once this has been done, it is a matter of assessing the factors one at a time and then cross-checking them for agreement (i.e., ensuring there are no contrary or paradoxical positions stated in different quadrants). Assessing can be done by asking hypothetical questions such as:

- In what way can the strengths be used to an advantage?;
- How can the weaknesses be shored up?;
- What is the best way to take advantage of each opportunity?; and
- What needs to be done to mitigate each threat?

In a tactical setting, analysts can use the results of SWOT to examine a target's operating structure, method of operating, capabilities, financial base, and so on. Strategies can be formulated based on combinations of the factors as follows:

Strengths/Opportunities. Ways that will use strengths so that opportunities can be realized.

Weaknesses/Opportunities. Ways to address weaknesses in order to provide relief so that opportunities can be followed.

Strengths/Threats. Ways that use strengths "offensively" to moderate threats.

Weaknesses/Threats. Defensive ways that will protect against threats.

PEST ANALYSIS

If an analytic tool could have a cousin, PEST could be said to be related to SWOT. PEST is an acronym for political, economic, social, and technological factors (social factors could be couched in slightly broader terms, such as sociocultural, if desired). These factors are usually the independent variables that are acting on the dependent variable. This technique has been used by the business community to assess the impact that these external factors might have on the organization or the market in which it operates. But, like SWOT, PEST can be used to an advantage by the intelligence community to assess a variety of issues under investigation.

Because PEST examines external factors, it is essentially half of a SWOT. But where the two differ is in the focus of the inquiry—PEST examines the environment in which the issue is positioned, whereas SWOT examines a dilemma or the actions of, say, a target. Viewed another way, PEST could be seen as the macro scene, and SWOT is the micro perspective. PEST is, therefore, used as the main tool for analysts when conducting what are termed *environmental scans*. In this sense, PEST is more likely to be used for strategic analysis where the issues are complex.

PEST analysis can be conducted before a SWOT analysis (e.g., via a workshop) to help identify issues. Though, it is less likely that a SWOT would be conducted before a PEST as there may be raised in a PEST that would subsequently feed into a SWOT.

There are many variations of PEST. Some analysts add additional factors, thus modifying the acronym to variants such as: STEP (PEST arranged differently—stay with PEST as it is universally known); PESTELI (adding environmental, legal, and industrial factors); PESTELO (same as PESTELI, but instead of the industrial factor, it uses organizational); STEEP (social, technological, economic, ethical, and political); and STEEPLED (social, technological, economic, ethical, political, legal, environmental, and demographic).

There is a view that some of these variants are not necessary and make the analysis overcomplicated. Some scholars have argued that the four factors under PEST cover all of the issues that would arise out of an examination of the other sub-factors. That is, the sub-factors are contained in the main PEST factors. Nevertheless, any of these factors can be mixed and matched to suit the research project.

Finally, like the SWOT, PEST has a simplicity that lends itself to ease of understanding while powerful enough to convey the results to decision makers. Two templates that can be used are shown in tables 10.2 and 10.3. One is a simple table, and the other contains more detail. They can be modified to suit most intelligence research projects or used according to the analyst's personal preference.

FORCE FIELD ANALYSIS

Force field analysis is a practical technique for examining the pressures that can be applied for or against a particular policy position, an operational tactic, or any other issue under investigation. It is a method that

Table 10.2. Simple PEST Analysis Template

POLITICAL	ECONOMIC	SOCIAL	TECHNOLOGICAL
• List Issues Here • A • B • C • D • E • F • G • H • I • J	• List Issues Here • A • B • C • D • E • F • G	• List Issues Here • A • B • C • D • E • F	• List Issues Here • A • B • C • D • E • F • G • H • I • J • K • L

allows the analyst to form a judgment based on careful weighing of the pros and cons involved.

For instance, a force field analysis can be carried out to weigh the possible success of a planned operation, or it can be used to gather *driving forces* in order to overcome or reduce the impact of *restraining forces*. Force field analysis also lends itself to other variations of possible use.

> "There is no 'perfect' decision. One always has to balance conflicting objectives, conflicting opinions, and conflicting priorities. The best decision is only an approximation—and a risk."[1]

Step by step:

1. In a few words, describe the issue, plan, proposal, or policy option in the middle of the diagram (see fig. 10.1);
2. In the left-hand column, list those attributes, options, or factors that can be considered driving forces in relation to the issue under investigation;
3. In the right-hand column, list those factors that are restraining forces;
4. Assign a numeric quantity to each force factor listed in the two columns. For instance, use an ordinal scale that ranges from, say, weak (+1) to strong (+5) for driving forces and for restraining forces, weak (−1) to strong (−5). In assigning a numeric quantity to each factor, analysts should be conscious that they do not bias the results by assigning values that reflect their own personal views, or the official view of the government of the day, or perhaps of one of the key decision makers

Table 10.3. Detailed PEST Analysis Template

	Comments and Observations	Impact Estimate High Medium Low Unknown	Timing 0–6 mths 7–12 mths 13–24 mths 24+ mths	Direction + Positive – Negative 0 Neutral	Rise/Fall > Increase < Decrease = Stable 0 Unknown	Import Critical Important Somewhat Not Very Not At All Unknown
Political	Include comments and observations here	High	6 mths	+	>	Not very
Economic	Ditto	Medium	12 mths	+	>	Not at all
Social	Ditto	Low	15 mths	–	<	Somewhat
Technological	Ditto	High	24 mths	0	=	Critical

(e.g., in order to curry favor). Projecting such bias into the method is unethical because it will artificially manipulate the analysis and use the scientific method simply as a guise for objective research. The safest way to assign the values is to achieve consensus through discussion with, for instance, a number of subject specialists that might take the form of a "judgment sample," "convenient sample" or other availability-based sampling technique.[2,3] If time and resources permit, the nominal group technique can be used to great advantage. Doing so removes any question of bias from the analysts (and the analytic unit that employs them).

5. Tally each column and add the two columns. If the total is a negative number, then the options regarding what the restraining forces are suggesting need to be considered carefully. If the number is positive, then the driving force options need to be considered. There is also the possibility that a zero result could occur or a weak (i.e., +1 or −1) result, suggesting that the direction may, on balance, be the way to proceed. But the analyst needs to apply judgment based on experience in such situations. Remember, these are just tools to guide thinking and reason; they do not reflect an absolute for any given situation.

As an example, consider this situation: an analyst is tasked with briefing decision makers about the likely impact of a newly proposed law that is intended to curb motorcycle gang violence. Based on the data obtained via the information collection plan, the analyst constructed a force field diagram similar to the one in figure 10.1.

Sources of information can also include brainstorming workgroups. This approach is particularly helpful if the timeframe of the issue being studied has a horizon of greater than six months. If the timeframe is measured in years, then complexity increases and convening a multidisciplinary (and, perhaps, multiagency) group to brainstorm the issues is a must.

Then, using a data projector, the analyst displays and talks about the force field diagram, discussing the pros and cons, and finally presents the agency's preferred position. This option will directly address the original research question (i.e., will the proposed new law have an impact on curbing motorcycle gang violence?).

When crafting recommendations, the analyst could also suggest changes to individual factors that, if implemented, could address the issue in favor of the driving forces, for it is the sum effect that the analyst is considering. For example,

- Suppose there was a restraining force like this: New anti-gang legislation could inadvertently contribute to the workload of the court system, which in turn would result in trial delays.
- A recommendation that adds force to the drivers could be crafted as such: To prevent over-listing and trial delays when the new anti-gang

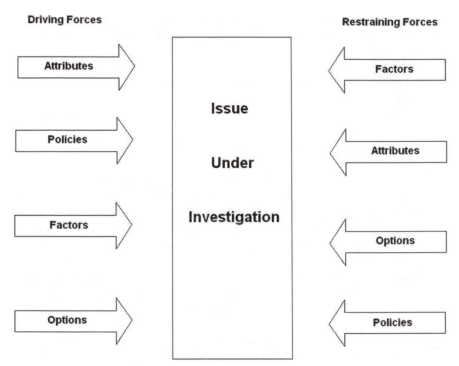

Driving Forces   **Restraining Forces**

Attributes   Factors

Policies   Attributes

Issue

Factors   **Under**   Options

Investigation

Options   Policies

Figure 10.1.   Force field analysis.

legislation is enacted, prosecutors would need to work with the criminal courts' listing coordinators to manage trial lists.

Recommendations formulated along the lines of the second item could tip the balance from a position that suggests restraining forces dominate to one that sees driving forces considered. Viewed another way, force field analysis is a way to identify not only the forces for or against the issue under investigation but also to create a chart that allows the analyst to visualize ways of countering restraining forces.

Pros-Cons-Fixes

Force field analysis is closely related to the decision making tool known as *pros-cons-fixes*. This variation starts by listing possible options for dealing with a problem. A corresponding note is placed next to each option, stating its positive attributes (this is usually done in table form). Then, all of the consequent negatives are listed, followed by ways of how the negatives can be overcome—these are the fixes. If no fix is available, then "no fix" is assigned to that option. The final step is to relist all of the options but omit those with fixes. This is because options that can be fixed are seen as neutral; they are extraneous and can confound judgment. What are left in the table are a number of pros and cons that can be assessed more clearly.

Plus-Minus-Interesting

A quantitative version of the pros-cons-fixes technique is termed *plus-minus-interesting*. The steps in conducting this type of analysis are the same, except that instead of labeling the column *fixes*, it is labeled *interesting* issues; these can be either positive or negative. In addition, each issue comment (for all plus, minus, and interesting issues) in the table is rated on a Likert-type scale ranging from –5 (very negative) to +5 (very positive), with 0 as the neutral point midway. Once this is done, the analyst adds the ratings for all issues listed to arrive at a score for each option. A separate table is constructed for each option under consideration so that the total derived for each option can be assessed against the others being considered. (See table 10.4.)

PARETO ANALYSIS

Pareto analysis is a quantitative way of prioritizing a list of options. The Pareto principle states that 80 percent of the effect of being observed is a result of 20 percent of the causes.

Pareto analysis, therefore, relies on a rating for each option. As such, Likert scores are often used. But whatever the metric, it must make sense in the context of the intelligence project. The following is a list of policy options generated by brainstorming; they relate to how a nation could deal with international pirates. Scores were assigned for each policy option after all of the ideas were written down. The higher the score, the more important the option was considered. Evaluating each option can be done through a

Table 10.4. Abbreviated Example of One Option of a P-M-I Analysis (Aggressive Stand against International Pirates)

Plus	Minus	Interesting
Boost domestic confidence (+3)	Some may see intervention as a risk to the lives of nationals living overseas (–2)	Media and press opportunities with other foreign leaders (+4)
Boost international reputation (+4)	Could be misinterpreted as a play for regional power (–2)	Joint operations (+2)
Secure safety of sea lanes (+5)	A further cost to the military (–1)	Valuable training for special forces (+2)
Ensure international trade (+3)	Military personnel to be overseas longer or more often (–1)	International law issues for intervention/interdiction (–3)

SWOT analysis or other evaluative process. Ideally, scores should be discrete, but if two options are considered equal, there is no reason why the same score cannot be assigned to each. Nevertheless, the final prioritized list must make sense to those who will be implementing the recommendations. In this example, the goal is to take an aggressive stand against international pirates. The six options are:

1. Increase surveillance by aircraft (+5);
2. Institute a system of in-country, covert "coast watchers" (+2);
3. Increase maritime naval patrols (+6);
4. Monitor radio communications frequencies used by pirates (+3);
5. Randomly search vessels of the class used by pirates (+4); and
6. Scuttle all vessels found carrying offensive arms or explosives (+1).

If this list is then prioritized according to the Pareto principle, an actionable list is created. From highest to lowest effect, the options are:

1. Increase maritime naval patrols;
2. Increase surveillance by aircraft;
3. Randomly search vessels of the class used by pirates;
4. Monitor radio communications frequencies used by pirates;
5. Institute a system of in-country, covert "coast watchers"; and
6. Scuttle all vessels found carrying offensive arms or explosives.

ANALYSIS OF COMPETING HYPOTHESES

Analysis of competing hypotheses is a useful tool to think about inductively construed theories in a rational way. It is an important tool when the analyst is faced with several plausible propositions to explain the issue under investigation. Rather than being compelled to accept just one theory, analysis of competing hypotheses allows the analyst to evaluate all theories.[4]

Through this technique, the available evidence suggests the most plausible theory rather than having to decide on subjective factors (here, the term *evidence* is a generic term that also applies to arguments and the like). If there are insufficient data to draw a conclusion, the techniques can aid a new (or revised) information collection plan so that further or better data can be fed into the process.

Managers of field operatives and other information collecting assets can use the output of this type of analysis to task their resources more efficiently, saving valuable time. Here are the steps involved in conducting an analysis of competing hypotheses:

- Draw up a matrix like the one shown in table 10.5, listing the various hypotheses across the top and important pieces of evidence (including a lack of evidence) down the left-hand column.[5] Each hypothesis and each piece of evidence does not have to be listed in full, a simple abbreviation of H1, H2 . . . and E1, E2 . . . will suffice. The analyst's descriptions in detail can be listed above or below the matrix as a reminder. At this point, the matrix is a quick way of bringing together all the information to form a clear picture.
- Working across the columns, assign a nominal value (+ or −) to indicate if each piece of evidence is consistent with the hypothesis or is inconsistent with the hypothesis.
- Tally the columns and consider whether the column with the most pluses should be advanced as the most likely hypothesis. The caveat placed on the conclusions drawn using force field analysis (see above) apply to this technique also.
- Just because one column may have the most "pluses" does not mean it *is* the best choice. Some pieces of evidence can and should carry more weight than others. In this regard, the process still requires the analyst to apply some degree of judgment before proceeding to advance this as the course of action.
- If sensitivity is an issue, analysts could consider using another scale, say, the ordinal scale, to attribute weight to each piece of evidence. For example, the use of double pluses (+ +) and double minuses (− −) could be added to the single signs discussed in this section, or a scale like that used in force field analysis could be adapted.

Although the use of the matrix is very useful, it is unlikely that any analyst will include it in the final report or briefing unless the audience is technically oriented (e.g., perhaps presenting the initial results to a peer group as part of a quality control process or validating the methodology).

Table 10.5. Competing Hypotheses Matrix: From Prunckun's Study of Heroin[6]

	H1	*H2*	*H3*	*H4*
E1	+	−	−	−
E2	−	+	−	−
E3	−	−	+	−
E4	−	−	−	+
E5	−	−	−	+
E6	−	−	−	+

Research Question

What are the likely factors that caused the heroin shortage in Sydney in 2001? The hypotheses are as follow (while the analysis is on the previous page, in table 10.5):

H1—Recent seizures (at that time) by law enforcement agencies.
H2—The arrest of significant personalities in the supply and distribution chain.
H3—A severe water drought in the poppy-growing regions of Myanmar (Burma).
H4—A Taliban-enforced reduction of Afghanistan-grown opium.

Evidence

E1—Quantitative data about Australian law enforcement seizures.
E2—Elimination of unnamed and unspecified personnel.
E3—Data on crop production and rainfall in Myanmar.
E4—Quantitative data on drug production in Afghanistan.
E5—A 3,000 metric ton reduction in Afghanistan-grown opium.
E6—Police intelligence of trafficking routes to Europe confirming a diversion of Golden Triangle heroin (destined for Australia) diverted to Europe to fill the Afghan void.

FISHBONE ANALYSIS

Fishbone analysis is used to show cause and effect by identifying and then exploring the surrounding problem under investigation. It can also be adapted by analysts as a tool to help manage the information collection process (see, for instance, fig. 2.1 in chapter 2).

The analyst lists the problem to be investigated at the right-hand side of the diagram (the fish's head). He or she constructs the major bones of the

Military intelligence analysts could use the nine order of battle factors to spotlight issues surrounding friendly, neutral, or enemy forces. These factors include: 1) composition of organizational units; 2) disposition (location and deployment) of units; 3) strength of units regarding personnel, weapons, and equipment; 4) tactics that would be used by the units; 5) training at individual and unit levels; 6) logistics for unit supply; 7) combat effectiveness of the unit; 8) electronic and communications technical capability; and 9) miscellaneous background and supporting information about the unit.

fish by listing the major categories of information concerning the issue. Traditionally, the categories are machinery/equipment, people, methods, and materials. Alternatively, the analyst can use categories such as policies, procedures, plant/equipment, and people. The categories used are not critical as they merely act as a framework for analysis.

From each of the major bones, minor bones can sprout to form a list of contributing issues. This is analogous to watercourses—small brooks and creeks that flow into larger streams, which, in turn, flow into rivers. Each issue contributes to form a larger problem. Fishbone analysis is aimed at identifying the problem's causes so that the analyst can suggest treatments. This analytic technique will not identify symptoms because the symptom is the focus of the analysis (i.e., the issue under investigation). See figure 10.2 for a template for conducting a fishbone analysis.

When populating the diagram, the analyst may find that some issues contribute to more than one problem. This is good to note because when seeking a solution, it might be advantageous to address those issues that contribute to more than one problem. Analysts can use methods such as brainstorming and the nominal group technique to generate ideas for treatment.

MORPHOLOGICAL ANALYSIS

The term *morphological* is used in the intelligence context to describe how analysts impose structure to their investigations in order to test multidimensional relationships. It is used to study how various concepts or ideas fit together. Morphological analysis is often used to explore complex policy issues where the factors involved are not quantifiable and would be difficult to model using statistical methods.

One of the strengths of morphological analysis is that it can generate a large number of possible explanations for a phenomenon. Or it can produce possible outcomes, treatment options, or causal theories for an event that has occurred or may occur in the future. In this regard, analysts can use the technique in place of brainstorming if time is a constraint or access to subject experts is not possible. Analysts can feed the output generated into subsequent analyses, such as competing hypotheses. Step by step:

1. Deconstruct the issue under investigation to expose its component parts. As an example, take the research question: what means are available for controlling cyber weapons? Headings for analyzing this problem could be borrowed from the fishbone analysis—that is, either the categories of machinery/equipment, people, methods, and materials or the categories of policies, procedures, plant/equipment, and people. Or analysts are free to devise their own categories.

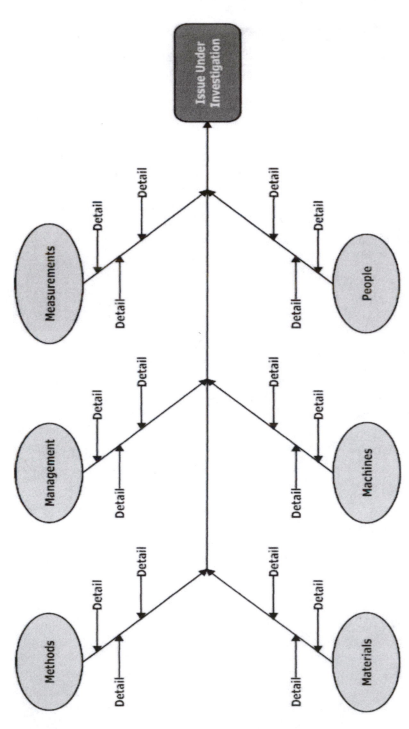

Figure 10.2. Typical template for conducting a fishbone analysis.

2. Create a matrix to organize the headings associated with the research question. Table 10.6 is an illustration taken from Prunckun's study of cyber weapons control.[7] In the row across the top, list the category headings that were identified in the deconstruction phase (step 1). In the columns extending downwards, list the possible elements that comprise the concept category heading (this order can be reversed, if desired, with the headings down the left-hand column and the possible elements across each row).

3. Generate possible scenarios. There are several ways to generate scenarios. Starting with the first heading, the analyst selects an element from somewhere in the corresponding column and then moves across one and selects an element from the next row, randomly. This is repeated until one element is selected from each row. Another, more methodical way, is to perform the selection in an orderly way that ensures every possible combination is included. In a two-by-six matrix this would be 36 possible outcomes. One can see that the matrix offers the analyst a mechanism for generating a large number of possibilities quickly.

4. Assess each of the explanations (i.e., output from step 3 above), and place in descending order from most plausible to least plausible. Some of the possibilities generated may be quite illogical; nevertheless, the process affords analysts a range of possibilities that they can now consider as a part of a set of action items for decision makers. Alternatively, analysts can use the process to just identify options (selectively) that they could not realize using other methods and, as such, can explore through further research.

PERCEPTION ASSESSMENT ANALYSIS

Perception assessment analysis allows the intelligence analyst to demonstrate relationships between actions taken by field operatives and the perceptions of those who will observe or experience those actions.

Table 10.6. Morphological Analysis Matrix

Prohibited	Regulated
Creation of cyber weapons	With existing criminal statutes
Possession of cyber weapons	With amendments to existing criminal statutes
Distribution of cyber weapons	Use existing licensing regime (e.g., for sporting guns)
Sale of cyber weapons	Create new licensing regime
Transfer of cyber weapons	For software weapons only
Use of cyber weapons	For hardware weapons only

In psychology and the cognitive sciences, perception is defined as an awareness derived though sensory information. This technique enables operational managers to understand the possible impediments that implementing certain actions might give rise to because of perceptions. A matrix format displays the analytic results so that decision makers can understand how perception may become a blockage.

Taking an example from the military, an assessment might show how others in the operational environment—"enemy, civilian population, multinational, or coalition partners"—could perceive the dealings they have with friendly forces.[8] The matrix is ideal for this as it lends itself to including other factors, including success criteria.

But in order to carry out this type of analysis, it requires a more than average degree of understanding about the social and cultural issues that dominate the nation, the region, or the locality where troops are operating (as there can be both subtle and noticeable changes from one area to another). Nevertheless, if the analysts do not possess this knowledge, there is no reason why they cannot glean this information from subject experts (e.g., through in-depth interviews or focus groups). This knowledge is then used to assess the likely reactions the observers might have to actions by friendly forces.

Measuring perceptions is a difficult science at best. One could argue that there is a relationship between the magnitude of the physical stimuli (i.e., say, actions taken by friendly forces) and how a person perceives these actions. But this may not be the case in a sociocultural setting—what friendly forces view as a positive activity, the local population could perceive as insulting or disrespectful. These actions could cause a backlash against the actions being taken by friendly forces and the forces themselves.

Analysts can measure perception by several methods. Four methods are suggested by the U.S. army,[9] and these are:

- Determine demographic and cultural factors that shape perceptions and reactions;
- Identify patterns and indicators from previous expectations and reactions in a society's history;
- Compare reported reactions to determine if they were based on real or perceived conditions; or
- Monitor editorial and opinion pieces of relevant newspapers for changes in tone or opinion shifts that can steer or may be reacting to the opinions of a society, organization, or group.

An example of a completed perception assessment analysis is shown in table 10.7. It appeared in the U.S. army's interim field manual entitled *Open Source Intelligence*[10] and shows across the top, the categories that

Table 10.7. Completed Perception Assessment Analysis

Condition	Cultural Norm	Friendly Force Action	Population Perception	Cause of Perception	Consequence if Unchanged
Food	Rice	Provided meat and potatoes	Inadequate and inconsiderate	Practical (no experience with potatoes) and cultural (dietary rules on meat)	Starvation and riots
Armed Civilian	All men carry weapons	Confiscated all weapons	Unfair and demeaning	Practical (safety) and cultural (symbol of manhood)	Risk of violence between US forces and armed civilians
Government Structure	Tribal	Established military administration (hierarchical	Tolerable as long as the authority fulfils needs	Historical (previous experience with Western or military forms of government)	Loss of credibility and eventually control if needs are not met

need to be considered before action is taken. In each of the rows are presented the issues relating to the corresponding heading. The first three columns at the right of center are essential factual data, but the three rows to the left of center are interpretations of the data to the right and, as such, require expert subject knowledge.

TIMELINES OF KEY DATES ANALYSIS

A sometimes undervalued analytic technique is timeline analysis. It is akin to descriptive statistical analysis in that it describes in qualitative terms issues associated with key dates. The findings are displayed in a chronology table or a modified version.[11] The table provides decision makers with a wealth of factual information that allows them to place proposed actions in sociocultural/political/religious contexts to avoid clashes with the local population.

Although what is discussed here is from a military perspective, timeline analyses can be used by law enforcers for issues like counterterrorism planning. These reports inform the decision maker how certain elements of the population might react to friendly force activity or when and why local law enforcement might expect an attack by terrorists.

These timelines list such events as national, regional, and local holidays as well as religious and cultural events. In areas that have had experienced political turmoil, public events and events that have symbolic political significance should also be noted.

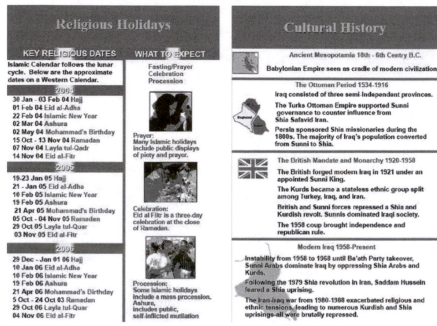

Figure 10.3.   Example of a timeline of key dates.

It is common to include descriptions of the demographic makeup of the population and political developments relevant to the planned operations. Information that features in reference works like *The CIA World Factbook*[12] could also be considered for inclusion in this type of analysis—geography, economy, people, government, communications, and defense forces. An example of a timeline of key dates is shown in figure 10.3.[13]

NETWORK ANALYSIS

If a research question involved the need to understand the relationships between two or more individuals, organizations, events, or other factors, then network analysis can help make these associations clear. The relationships can be anything—social, business, financial, or even relationships that show abstract concepts such as influence, support, or mentoring.

The origin of network analysis is in the social sciences where scholars, like Moreno,[14] devised the use of two dimensional diagrams to display relationships.[15] These were, and are, called *sociograms*, but in intelligence work analysts have termed them *network analysis*.[16] Network analysis should not be confused with the closely related analytic technique known as *traffic analysis*.

Network analysis has a common bond with *traffic analysis*; the latter involves intercepting radio or telecommunications traffic between entities. Using analytic tools, patterns in these communications can reveal inferred meaning in the context of the issue under investigation. It is the exchange of the communication that is the subject of the analysis, not the content of the message. Times, days, frequency, duration, method of transmission, encryption method, and so on are the elements that are studied. As such, traffic analysis is an important methodology for situations where the targets are using encrypted radio transmissions.

Network analysis is also called *association analysis* and *link analysis* because during the process, analysts use a matrix to show associations and lines to shown links.[17] Although these terms are used, this author is in favor of standardization with the term *network analysis* as the other terms only describe part of the overall analytic technique. Because network analysis has a long history and tradition within the intelligence community,[18] analysts should stay with this term and avoid others.

Network analysis consists of plotting nodes that represent entities—circles indicate people, squares indicate companies or businesses, solid lines represent strong associations, dotted lines are used for weak or unconfirmed links, and the numbers along the lines count the number of contacts each entity has had with the other during the period of the study. Concepts such as *influence* can be represented by using arrows instead of lines.

The individual links that comprise the overall network can be described by the attributes of the associations between the entities. In a criminal context, these attributes might include routine pieces of information, like victims' addresses and phone numbers, their movements prior to the alleged offenses, and the offenders' modus operandi—all of which are analyzed in order to generate investigative leads, infer an organization's hierarchy (or lack of one), determine points of vulnerability/strength, and so on. Step by step:

1. Identify all entities. This can be done through other forms of analysis such as telephone record analysis or distilled from surveillance reports, notes or transcripts of interviews, telephone or email intercepts, documents seized during raids, and so forth.
2. Assemble these data in an association matrix (see table 10.8 for an illustration). This is a standard matrix consisting of the same entities listed across the top and down the left-hand side. Where an entity intersects itself, the cell is blanked out. Where there is a known or

confirmed association between two entities, a solid dot is entered into the corresponding cell for the two. If the association is suspected, a hollow dot is used, and a plus sign is used to denote that a person is a key individual in a company or organization.

3. Depict the entities as symbols on a chart (the chart can be a marker board, a flipchart, or a computer document), and draw the relationship lines (i.e., links) between them (see fig. 10.4). Common symbols are: circles for people, squares for corporate entities, and circles within squares for persons associated with an organization; solid lines represent strong associations, dotted lines for weak or unconfirmed links. Influential relationships use arrows pointing from dominant to subordinate. If using a computer program, it will have preassigned symbols for entities, for instance a telephone symbol for telephone numbers, a boat for watercraft, a car or truck for motor vehicles, and so on. There is no correct way to display the entities on the chart. As intelligence projects differ, so too will each chart. The number and relationship between the

Table 10.8.   Example of an Association Matrix

	Jack's restaurant	Jack	Rosa	Liz	Jerry	Dom
Jack's restaurant						
Jack	+					
Rosa	+	+				
Liz	+	+	+			
Jerry				+		
Dom	+	+	+			

entities will vary. As such, analysts will have to use their sense of artistic arrangement to create an exhibit that will clearly and easily show the viewer the relationships. If it is too "busy," it is likely to be confusing. If it is confusing, it defeats the purpose of presenting the data in this way.

TELEPHONE RECORD ANALYSIS

This is a variation to the association matrix discussed in network analysis above. This technique is performed in the same way as association analysis, but instead of listing entities, the matrix lists the telephone numbers that were called on the left-hand side and the numbers that made the calls across the top. Where associations exist, a numeral is used to record the number of times that telephone called the other.

Taking these data, analysts construct a chart exactly as they did for network analysis, but arrows are used to show the numbers making the calls to other numbers. The frequency of calls made is shown as a numeral alongside the arrow. Landline telephones, pagers, mobile (cell)

Figure 10.4.   Example of a network chart showing links.

telephones, fax machines, and computer modems can all be part of this analysis.[19]

The technique streamlines an otherwise labor-intensive process, especially if there is a large volume of data (e.g., obtained from telephone bills or pen registers). If this technique is used in conjunction with a network analysis of entities, the findings could help identify targets for installing covert listening devices or for conducting physical surveillance.

EVENT AND COMMODITY FLOW ANALYSIS

Event flow analysis is a technique that is used to clarify situations that comprise multiple events and take place over a period of time—hours, days, weeks, months. These situations are usually characterized by complex simultaneous or closely related historic events that would be confusing to understand unless ordered sequentially in time.

Analysts can also use this technique to understand the flow of commodities—illicit drugs, stolen goods, importation or exporting of restricted/prohibited items, sales of weapons or their components, explosives, precursor chemicals, technology, and the like.

An event/commodity flow analysis is a diagrammatic depiction of the chronological events/movements. Once visualized in this way, analysts can draw inferences to generate further investigative leads or to test a hypothesis. Step by step:

1. Identify all events associated with the issue under investigation. Analysts can obtain these data from crime scene examinations or, as with network analysis, through the distillation of surveillance reports, notes or transcripts of interviews, telephone intercepts, documents seized during raids, and so on.
2. Assemble these data in a chronology table. This is a similar process to the timeline table produced in the analysis of key dates (above).
3. Depict each event as a symbol on a chart (the chart can be a marker board, flipchart, or computer document), and draw a progress line between them. Times that separate the events can be shown as vertical dividing lines at uniform time periods. If two or more events occur at the same time, then the progress lines joining the events will split and connect to the simultaneous events. These lines will converge again as they move to the next event(s).

GENEALOGICAL ANALYSIS

Genealogy is the study of families by searching their lineages back through history. From time to time, intelligence analysts may be called upon to provide a descriptive analysis of the extent of a target's family relationships. A case in point is the late dictator of Iraq, Saddam Hussein, where a detailed understanding of his family's relationships aided in his capture in December 2003.[20]

These data come from many sources: oral histories (of relatives); personal historical records (e.g., family bibles); birth, death, and marriage records and newspaper notices of same; grave and burial records; census records; and military and other government records. Analysts can use any source of information that shows kinship or pedigrees. Contrast genealogy with family history research—the former is a study of kinship, while the latter is a more detailed investigation into the lives and history of the family members.

Analysts display genealogical research findings in chart form as the visual presentation of the associations is easy to follow. They also use written narratives, but unless the number of kinships is few, such narratives can confuse the reader (though analysts can use an indented outline to show generations more plainly).

Analysts can draw pedigree charts in a number of ways, including freehand and with preprinted templates. But the use of a computer package to collate and display the relationships is the most efficient method. These packages typically have other features for producing reports, including the ability to generate: group sheets for individual families, descendent or ancestor charts, individual relationship charts and individual summaries, lists of anniversaries (e.g., birth, death, marriage), and if photographs of the family members are available, these can be incorporated into the genealogical database and displayed in the pedigree chart as illustrated in the invented example shown in figure 10.5.

FINANCIAL ANALYSIS

The single most useful analytic technique for the non-account analyst is net worth analysis. In most agencies, financial analysis is conducted by a qualified accountant because of the degree of knowledge it requires to understand the processes and procedures to reconstruct the target's financial position.

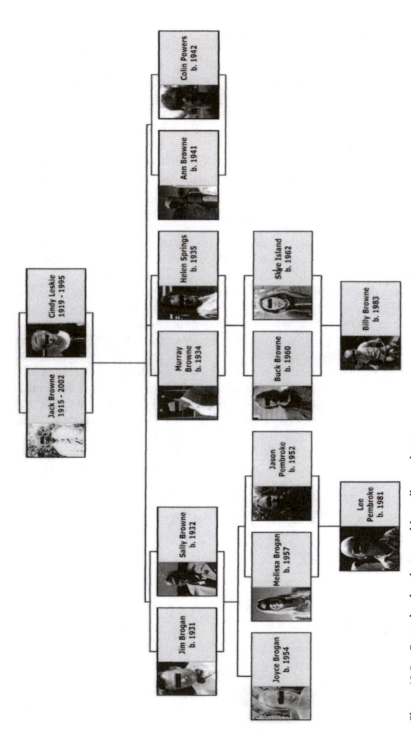

Figure 10.5.   Example of a photographic pedigree chart.

Nonetheless, an intelligence analyst can conduct a preliminary analysis of the target's net worth using the methods described here. If the analogy of a paramedic and a surgeon is used, the intelligence analyst would be the paramedic performing the first stage analysis while calling in the surgeon to perform the more intricate analysis once the major issues are identified.

Net Worth Method

Net worth is an indirect method of assessing the target's income and, therefore, is a handy tool in situations where the analyst may have come across information that suggests some form of illegal enterprise. It could also be used to assess a target's suitability to an approach by a field operative who is hoping to recruit the target as an agent (e.g., an offer of financial assistance).

Net worth is simply the difference between the target's assets and liabilities. If the analyst conducts a net worth analysis over a period of time, say, for the end of each financial year, he or she can compile a picture as to whether the target is growing in worth or is experiencing losses and what the magnitude of these gains or losses might be. The formulas for calculating net worth are as follows, step by step:

1. Assets – liabilities = net worth
2. Net worth – prior year's net worth = increase or decrease in net worth
3. Net worth increase (or decrease) + living expenses = income
4. Income – funds in known sources = funds from potentially illegal sources[21]

KEY WORDS AND PHRASES

The key words and phrases associated with this chapter are:

- Analytic tools and techniques;
- Association analysis;
- Competing hypotheses;
- Driving forces;
- End-state;
- Environmental scan;
- Event and commodity flow analysis;
- Evidence;
- Financial analysis;

- Fishbone analysis;
- Force field analysis;
- Genealogical analysis;
- Link analysis;
- Morphological analysis;
- Network analysis;
- Net worth method;
- Pareto principle;
- Pedigree chart;
- Perception assessment matrix;
- PEST analysis;
- Restraining forces;
- Sociogram;
- SWOT analysis;
- Telephone record analysis; and
- Traffic analysis.

STUDY QUESTIONS

1. What are the four quadrants in a SWOT analysis? Describe each, and explain what type of data the analyst would seek to populate each.
2. Compare and contrast SWOT with PEST. Discuss when an analyst might use one technique over the other and why.
3. Describe the variations that can be applied to PEST.
4. Summarize the steps in creating a network analysis.
5. Discuss situations when an analyst might use event flow analysis and commodity flow analysis.

LEARNING ACTIVITY

Suppose you have been tasked to construct a pedigree chart for the dictatorial leader of Country Q. Your findings will form part of a psycholinguistic analysis of him. Using hand drawn lines or a software package, create a pedigree chart. For the purpose of this learning activity, use your own family members as a way of indicating your skill in creating the chart. Seek as many different sources of data as possible to simulate a real-world project. If stumbling blocks are encountered in obtaining data about family members (as they would be with a real target), consider how you could collect this information from other sources. List the possible alternative sources and methods of acquiring these data (e.g., using an information collection plan format).

NOTES

1. Peter F. Drucker, *Management: Tasks, Responsibilities, Practices* (Woburn, MA: Butterworth-Heinemann, 1974), 387.

2. Gary T. Henry, *Practical Sampling* (Newbury Park, CA: Sage, 1990), 17–20.

3. Gennaro F. Vito, Edward J. Latessa, and Deborah G. Wilson, *Introduction to Criminal Justice Research Methods* (Springfield, IL: Charles C Thomas, 1988), 127–28.

4. Richards J. Heuer, *Psychology of Intelligence Analysis* (Washington, DC: Center for the Study of Intelligence, Central Intelligence Agency, 1999).

5. Hank Prunckun, "A Rush to Judgment?: The Origin of the 2001 Australian 'Heroin Drought' and Its Implications for the Future of Drug Law Enforcement," *Global Crime* 7, no. 2 (May 2006): 247–55.

6. Prunckun, "A Rush to Judgment?", 247–55.

7. Hank Prunckun, "'Bogies in the Wire': Is There a Need for Legislative Control of Cyber Weapons?" *Global Crime* 9, no. 3 (August 2008): 262–72.

8. U.S. Department of the Army, *FMI 2-22.9: Open Source Intelligence* (Fort Huachuca, AZ: Department of the Army, 2006), 4–16.

9. U.S. Department of the Army, *FMI 2-22.9*, 4–16.

10. U.S. Department of the Army, *FMI 2-22.9*, 4–16.

11. Marilyn Peterson, *Applications in Criminal Intelligence: A Sourcebook* (Westport, CT: Greenwood Press, 1994), 36.

12. Central Intelligence Agency, *The CIA World Factbook, 2009* (New York: Skyhorse Publishing, 2008).

13. U.S. Department of the Army, *FMI 2-22.9*, 4–18.

14. Jacob L. Moreno, *Who Shall Survive? Foundations of Sociometry, Group Psychotherapy, and Sociodrama* (Washington DC: Nervous and Mental Disease Publishing, 1934).

15. John Scott, *Social Network Analysis: A Handbook* (Newbury Park, CA: Sage, 1991).

16. Henry Prunckun, "The Intelligence Analyst as Social Scientist: A Comparison of Research Methods," *Police Studies* 19, no. 3 (1996): 67–80.

17. International Association of Law Enforcement Intelligence Analysts, *Successful Law Enforcement Using Analytic Methods* (Alexandria, VA: IALEIA, 1997).

18. Francis Ianni and Elizabeth Reuss-Ianni, "Network Analysis," in Paul Andrews and Marilyn Peterson, eds., *Criminal Intelligence Analysis* (Loomis, CA: Palmer Enterprises, 1990).

19. Marilyn Peterson, "The Context of Analysis," in Paul Andrews and Marilyn Peterson, eds., *Criminal Intelligence Analysis* (Loomis, CA: Palmer Enterprises, 1990).

20. U.S. Department of the Army, *US Army Field Manual 3-24/ Marine Corps Warfighting Publication 3-33.5* (University of Chicago Press, 2007), 323.

21. Leigh Edwards Somers, *Economic Crimes: Investigating Principles and Techniques* (New York: Clark Boardman Company, 1984), 99.

11

✺

Analytic Techniques for Counterterrorism

This chapter discusses the analytic techniques that are used in counterterrorism analysis:

1. Threat analysis;
2. Vulnerability analysis;
3. Risk analysis; and
4. Prevention, preparation, response, and recovery planning.

INTRODUCTION

Threat analysis is the first of three integrated phases in developing a counterterrorism plan. The two subsequent phases are vulnerability analysis and risk analysis. The results of these three pieces of analytic work lay the groundwork for crafting a policy that addresses *prevention, preparation, response,* and *recovery* (PPRR). In other words, all of the techniques contained within this chapter are intrinsically linked and act as building blocks to form a comprehensive "toolkit" for analyzing counterterrorism. These steps in summary form are:

1. Identify the threat(s);
2. Explore vulnerabilities to this threat(s);
3. Gauge the likelihood that the threat(s) will eventuate;
4. Assess the consequence the threat will have; and
5. Construct a PPRR plan.

Consider the following example of how these steps are applied in practice:

1. Threat—cyber attack via an email-borne virus;
2. Vulnerability—the agency's servers and workstations via the Internet;
3. Likelihood—greater than 85 percent probability;
4. Consequence—moderate to severe loss of computer resources; and
5. PPRR—develop a plan that does four things: attempts to prevent such an attack (prevention); prepares the agency for such an attack if prevention measures fail (preparation); guides the agency in the actions it needs to take to respond to an attack that is underway or has occurred (response); and, suggests what needs to be done to aid the agency in recovery once an attack has passed (recovery).

Although discussed here as a packaged approach to counterterrorism, any one of these analyses can be carried out on its own or applied to problems other than terrorism. For instance, a risk assessment could be conducted in relation to a person or group acting criminally.

The term *counterterrorism* has been adopted in widespread use for defensive measures over the term *antiterrorism*. Strictly speaking counterterrorism involves offensive measures taken to "prevent, deter, pre-empt, and respond to terrorism."[1] Nonetheless, counterterrorism is used in this text as it is in common use by many law enforcement and security agencies worldwide.

THREAT ANALYSIS

A *threat* is a person's resolve to inflict harm on another. Threats can be made against most entities—people, organizations, and nations (i.e., by a *threat agent*). The potential harm can be in many forms and can be suffered either physically or emotionally/mentally. Threat agents do not have to openly declare their resolve to cause harm in order to constitute a threat, though explicit words or actions make it easier for field operatives to identify the threat agent and for analysts to assess the threat.

Threat analysis acknowledges two key factors—that there needs to be a threat agent (which could be anything from a physical substance to a person or a body corporate/organization) and an object of the threat (i.e., the target—which does not have to be a material target such as a shopping mall or an individual—can be intangible such as the threat to national security or the security of a particular venue or event). Stated another way, a threat agent who has intent and capability must be able to harm something. By way of example, a threat agent could be a drug trafficker who is intent and capable of illegally importing, say, heroin or a group of insurgents who has an intent and capability to destroy a bridge, and so on.

When analysts assess a threat agent, they are gauging whether the agent has *intent* and *capability* to produce harm to a target. To weigh whether the agent has intent and capability, analysts need to establish two elements for each of these factors: *desire* and *expectation* (or *ability*) for intent, and *knowledge* and *resources* for capability. These considerations are shown diagrammatically in figure 11.1. As an equation, threat is expressed as:

$$threat = (desire + expectation) + (knowledge + resources)$$

Desire can be described as the threat agents' enthusiasm to cause harm in pursuit of their goal. Expectation is the confidence the threat agents have in that they will achieve their goal if their plan is carried out. Knowledge is having information that will allow the threat agents to use or construct devices or carry out processes that are necessary for achieving their goal. Resources include skills (or experience) and materials needed to act on their plan.

> *threat intent = the optimism a threat agent has about successfully attacking a target*
>
> *threat capability = the force a threat agent can bring to bear on a target*

A fishbone analysis could be used to show the factors that contribute to each of these elements in a cause-and-effect relationship (refer to fig. 10.2 in chapter 10). Recall from our discussion of fishbone analysis in chapter 10 that at the fish's head is listed the problem to be investigated—in this case, it is "threat." The major bones of the fish constitute the significant categories of information concerning the problem—desire, expectation, knowledge, and

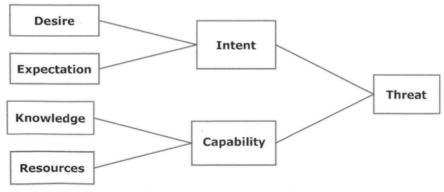

Figure 11.1.   Threat analysis.

resources. From each of these major bones, minor bones sprout, comprising the contributing factors that constitute each of the four categories.

Analysts need to consider the context of the threat, their agency's mission, and the list of potential targets when adopting a model in order to aid them in determining the threat environment. A model for calculating threats might look something like table 11.1.

Though models do not eliminate subjectivity, using a model forces the analysts to be transparent about how they calculate threat and, in doing so, are able to defend their conclusions. You will note that there is no weighting attached to what constitutes a high level of intent. That is because one cannot say how many media announcements it would take from, say, al-Qaeda to represent high intent. Ideally, some form of

Table 11.1. Threat Posed to the Orrenabad Community by the Omen Martyrs Faction

Scale	Scores	Tally
Desire		
Negligible	1	
Minimum	2	
Medium	3	3
High	4	
Acute	5	
Expectation		
Negligible	1	
Minimum	2	
Medium	3	3
High	4	
Acute	5	
Intent		6
Knowledge		
Negligible	1	
Minimum	2	
Medium	3	3
High	4	
Acute	5	
Resources		
Negligible	1	
Minimum	2	
Medium	3	3
High	4	
Acute	5	
Capability		6
Threat coefficient		12

conditioning statement would be attached to each of these categories so that the decision maker knows what is meant by high intent, low intent, and so forth. An example of how such a conditioning statement scale could be constructed is shown later in this chapter, in table 11.5.

In addition, models do not eliminate miscalculations because of inadvertent skewing. Note in table 11.1 that intent is calculated by adding desire with expectation, and in turn, this sum is added to the sum of knowledge and resources (and will range from a low of 4 to the maximum of 20). The process of adding limits the spread of values, whereas the process of multiplying any of these scores would increase the values. For instance, if all scores were multiplied—that is, substituting multiplication for addition—as per the equation, the range would be spread from 1 to 625.

The precision of this wide range of values diminishes the analyst's ability to accurately determine either intent or capability. Therefore, it is suggested that adding all values, rather than multiplying them, will reduce the spread and, therefore, maintain the threat coefficient as an *indicator* rather than promote it as a reflection of its absolute condition. (Even if the analyst multiplied desire and expectation, and knowledge and resources but added the resulting sums, it would still yield a very wide spread—from 2 to 50—as would the opposite, that is, multiplying the sums that comprise intent and capability, from 4 to 100.)

Having said that, two additional issues need to be noted: 1) there is still a need to provide conditioning statements so that the reader of the intelligence report understands what is meant by a medium threat intent and capability (e.g., along the lines of table 11.5), and 2) "unknowns" are not accommodated in this model.

The threat coefficient obtained from this analysis is then compared against a reference table to gauge where it sits on the continuum of danger of attack. The scale suggested in table 11.2 can be varied with additional qualifiers, or it can be collapsed if the number is deemed too many. Likewise, how the incremental breakdown of coefficients is determined will depend on whether the agency is willing to accept the risk that a threat agent may slip under its gaze by raising the categories of negligible and minimum. In the end, the number and their descriptors need to make sense in the context of the asset being protected. That is, each of the descriptors needs to have a conditioning statement attached to it to define

Table 11.2.   An Example of a Threat Coefficient Scale

Threat	Coefficient
Negligible	4–6
Minimum	7–10
Medium	11–15
High	16–18
Acute	19–20

what is meant by negligible, minimum, medium, high, and acute. Table 11.5 (shown later in this chapter) is an example.

Threats are context dependent, and what forms a threat in a business setting does not necessarily form a threat in a military setting or national security setting (though the opposite may be true). Bearing this in mind, an example from the military will be discussed to illustrate the threat analysis method.

In a low intensity conflict, threats can range from spontaneous street demonstrations by the local population at one end to terrorist bombings and confrontations with insurgent or guerrilla units at the other end. The techniques for assessing the elements of a threat can vary depending on the issue under investigation and the analyst's personal preference or the agency's policy.

Nevertheless, the approach is to weight each element using some verifiable means that is open to third-party scrutiny. For instance, an analyst may use a force field analysis to judge whether there are threats in Country Q associated with a low intensity campaign being prosecuted by friendly military units. Likewise, the nominal group technique could be employed not only to assess the four elements of a threat (i.e., desire, expectation, knowledge, and resources) but also to generate a list of possible threat agents (i.e., belligerents) to compare the elements against each other. Participants for such a group could be drawn from subject experts or operational specialists or a mixture of both. Some of the other analytic techniques discussed in chapter 10 can also be used. There is no firm rule on how this analysis should be done.

One way of contextualizing threats is to see them as *threat communities*. Some examples of threat communities pertaining to malicious human threats include:

External

- Competitors;
- Criminals and criminal groups;
- International or transnational terrorists;
- Insurgents and guerrillas;
- Domestic anarchists;
- Cyber law breakers;
- Rights campaigners;
- Spies-for-hire (i.e., ex–law enforcement, security, or intelligence personnel who have turned private operatives); and
- Foreign government intelligence services.

Internal

- Principals of the business or corporation;
- Associates;

- Employees; and
- Contractors.

These threat communities can be subdivided into more distinct groups if there is a need—for instance, rights campaigners can be classified into political activists, religious activists, single-issue activists (anti-whaling, animal rights, anti-abortion, environmental, etc.). But bear in mind that membership of one threat community (or subcommunity) does not exclude that person being a member of another or several other threat communities.

> When compiling a threat profile, targets can and should be considered in terms of their criticality, cost (either as a direct loss or an indirect or consequential loss due to disruption), or sensitivity (e.g., compromised information). This is because targets that do not possess any of these attributes may not be considered by threat agents with the same weight.

To better understand the "who" that comprise a threat community, analysts need to compile a *threat profile*. The profile needs to be adequate (perfection is rarely, if ever, obtainable) in order to understand the threat environment, which aids the next phase in counterterrorism analysis—that is, vulnerability analysis. In the meantime, consider the threat profile shown in table 11.3 as an example that demonstrates the important aspects of a fictitious threat agent (the order can be rearranged to suit the analyst's research project, and other factors can be added if these are deemed inadequate to communicate the message).

VULNERABILITY ANALYSIS

In short, *vulnerability* is a weakness in an *asset* that can be exploited by a threat agent (the term *asset* is being used in this context to denote a resource that requires protection). Viewed another way, vulnerability is an asset's capability to withstand harm inflected by a threat. Harm can be anything from experiencing a minor nuisance event to a situation that is catastrophic.

Vulnerability is a function of several factors—attractiveness of the target, feasibility of carrying out an attack, and potential impact. This model is shown diagrammatically in figure 11.2. Usually, these factors entail such considerations as: status of the target, potential for the attack to succeed, potential for the threat agent to get away with the attack, and potential to inflict loss. These factors can be weighed against measures to mitigate loss and to deter or prevent attack on an asset (e.g., through a force field analysis).

Table 11.3. Threat Profile for the Notional Omen Martyrs Faction

Summary Type	Observations
Organization	
Organization	Well organized but not hierarchical
Affiliation	Autonomous
Recruitment	Ethnic population centers
Financing	Extortion and kidnapping the wealthy
International connections	Training and ideological support
Behavioral	
Motivation	Radical religious ideology
Intent	Extensive destruction
Tolerance to risk	High
Self-sacrifice	Very accepting
Willingness to inflict collateral harm	Extreme
Operational	
Planning	Based on target acquisition intelligence
Targets	Objects that represent Western values or people who do not ascribe to their interpretation of their faith (including other believers)
Target characteristics	Symbolic and iconic objects that afford high visibility and, hence, high media coverage
Tactics	Targets mass gathering, critical infrastructure, communications, mass transport, and distribution chains
Weapons	Improvised explosives and small arms
Resource summary	
Skills and knowledge	Attack vector dependent (e.g., computer—low; electronic—moderate; explosive—high)

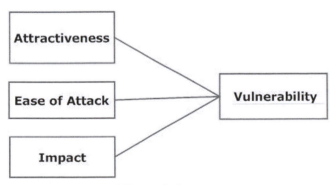

Figure 11.2. Vulnerability analysis.

Formulae-based analyses are popular among law enforcement and security agencies engaged in counterterrorism, and although these vary from agency to agency, they follow a basic stepwise formula:

1. Define what constitutes an asset (critical infrastructure, transport network, food chain, distribution hubs, or any of the essential services—e.g., electric power, gas, potable water, sewerage, etc.);
2. Sort these assets into categories;
3. Assign a grade or level of importance to each asset; and
4. Identify potential impact on the asset if it suffers harm.

As there is no one single criterion for calculating vulnerability because each class of asset may require special considerations to be taken into account (and there may be agency protocols that take precedence also), one general approach is to use a model such as:

$$vulnerability = target\ attractiveness + ease\ of\ attack + impact$$

To operationalize *attractiveness*, the analyst could ask questions along the following lines and tabulate the results to insert into the model:

- Is the target readily recognizable? Rather than answer this question in a dichotomous way (i.e., using nominal data—yes/no), the analyst could use ordinal data to give greater precision to the overall vulnerability indicator, for example: is the target recognizable internationally in the same way the World Trade Center was, or is it recognizable only nationally, statewide, or just locally? (See fig. 11.3.)
- Is the target the subject of media attention/coverage? Again, an analyst could construct a scale of attention from rarely to frequently/weekly. Coverage could be in the local press or through to global newscasters.
- Does the target have a symbolic status in terms of historical, cultural, religious, or other importance? The analyst could assess this factor as having no symbolic status to having multiple imports.

Attractiveness needs to be placed in context with the threat agent. Say, for instance, some Islamic extremist groups may see assets that represent Western culture or symbolize Western values as attractive.[2] To operationalize the concept *ease of attack*, the analyst could ask these types of questions:

- How difficult would it be for the threat agent to predict the peak attendance times at the target? Establish a scale from certain (as in the case of published opening hours) to very difficult (in the case of a training center located in a remote area and opened only for ad hoc lectures).
- Are there security measures in place (e.g., calculated on a scale of low to high deterrence or low to high prevention)?

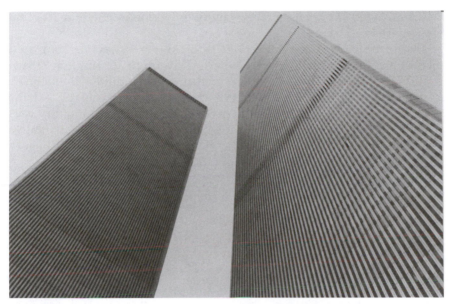

Figure 11.3.   International recognition: The former World Trade Center.

Questions that probe the existence and extent of controls (or lack thereof) can also be asked to gauge ease of attack. On the one hand, if there is a high degree of control effectiveness, this will usually reduce ease. On the other hand, if there is a low level of control effectiveness, it will increase ease. Analysts should be mindful that, with some targets, even a small reduction in control effectiveness can result in a disproportional increase in ease of attack. *Impact* could be operationalized by questions like:

- What are the numbers of people frequenting the target? Establish a scale ranging from a few daily/weekly/monthly to hundreds or thousands daily/weekly/monthly. Are these same people attracted from the local community, or are they international tourists?
- In dollar terms, what would the financial impact of an attack be if the asset was: disrupted, incapacitated, or destroyed? Or it could be put in terms of hours without operation, units of production, and so on.

Impact is predicated on an assumption that terrorists want their attacks to result in large numbers of deaths. This may be true in the al-Qaeda focused climate that existed at the time of this writing, but such an assumption may not always be valid; for instance, there may exist a nationalist-focused group that seeks to destroy infrastructure rather than kill people. In such a case, these terrorists may view heavy public traffic as an inhibitor to ease of attack. The two paradigms could be described as *effect-based attacks* versus *event-based attacks*.[3]

A template for calculating vulnerability might look something like that shown in table 11.4.

The vulnerability coefficient derived from this analysis is then compared against a reference table to gauge where it sits on the continuum of susceptibility to attack. The scale can be increased with additional qualifiers, or it could be collapsed if the number is deemed too many. In the end, the number and the descriptors need to make sense in the context of the asset being protected (the left-hand and center column of table 11.5). Qualitative descriptors (i.e., conditioning statements) can be added for each category as shown in the right-hand column of table 11.5.

Note that *consequence* is not a factor that is considered in a threat assessment. It is, however, considered in a risk assessment (see the following section).

RISK ANALYSIS

Risk is a function of *likelihood* and *consequence*. (The term *probability* is sometimes used instead of *likelihood*. Both are acceptable.) A risk assessment can be carried out in relation to almost any situation; it is not just

Table 11.4. Vulnerability of the City's Main Bridge over the Orrenabad River

Scale	Scores	Tally
	Attractiveness	
Negligible	1	
Minimum	2	
Medium	3	3
High	4	
Acute	5	
	Ease of Attack	
Negligible	1	
Minimum	2	
Medium	3	3
High	4	
Acute	5	
	Impact	
Negligible	1	
Minimum	2	
Medium	3	3
High	4	
Acute	5	
Vulnerability coefficient		**9**

Table 11.5. Vulnerability Coefficients

Vulnerability	Coefficient	Qualifier
Negligible	1–3	• Can only be attacked successfully if the threat agent has an acute threat coefficient; or • Has little or no symbolism; or • The range of security measures makes attack very difficult; or • If attacked, the target has significant redundancy and/or has sufficient capacity to continue functioning.
Minimum	4–6	• Can only be attacked successfully if the threat agent has a high coefficient (or greater); or • Is symbolic locally or regionally; or • The range of security measures makes attack difficult; or • If attacked, the target has a large measure of redundancy and/or has a large capacity to continue functioning.
Medium	7–9	• Can be successfully attacked if the threat agent has a medium coefficient (or greater); or • Is a well recognized regional symbol and/or a somewhat recognized national symbol; or • The range of security measures makes attack moderately difficult; or • If attacked, the target has a moderate amount of redundancy and/or has an average capacity to continue functioning.
High	10–12	• Can only be successfully attacked if the threat agent has a minimum threat coefficient (or greater); or • Is a well recognized national symbol; or • The range of security measures makes attack undemanding; or • If attacked, the target has little redundancy and/or has little capacity to continue functioning.
Acute	13–15	• Can only be successfully attacked if the threat agent has a low threat coefficient (or greater); or • Is a well recognized international symbol; or • The range of security measures is non-existing; or • If attacked, the target has no redundancy and/or no capacity to continue functioning.

for issues of grave concern. Nor is risk management solely for counter-terrorism; risk analysis techniques can be applied to situations that may be the target of criminals or criminal organizations not associated with terrorism. Nevertheless, analyzing risk allows analysts to recommend measures that will provide field commanders with the ability to:

- Accept the risk as is; or
- Treat the risk (which includes such decisions as to avoid the risk altogether, mitigate the risk, or defer the risk to another person or agency).

In intelligence research, analysts can focus on a wide range of risks. These can vary from the minor—say, a noncritical facility—to risks that liberal democratic nations face from the likes of weak and corrupt governments, rogue states, sub-state, and trans-state actors as well as organized criminal, radical ethic, racial and religious groups, and ultra right-wing political groups.

Internationally, risk analysis is the subject of a standard. The Swiss-based International Organization for Standardization (ISO) has published a document that puts forward a common approach for dealing with risk by providing generic guidelines in relation to the principles for how risk is managed.[4] In Australia, as well as in New Zealand, uniformity in risk management is specified by AS/NZS 4360:2004. This document is published through a joint venture by these two organizations: Standards Australia and Standards New Zealand. AS/NZ 46300:2004, along with the ISO 31000, can be applied to a number of activities, decisions, or operations in the private and public sectors as well as the military. They can also be applied by nonprofit organizations, community groups, and individuals.

Some of the key terms used in risk management are the name of the technique considered here—*risk analysis*—as well as *risk, risk assessment,* and *risk management*. According to AS/NZ 4360:2004,[5] risk is "the chance of something happening that will have an impact on objectives." Risk assessment is "the overall process of risk analysis and risk evaluation," and risk management is "the culture, processes and structures that are directed towards realizing potential opportunities whilst managing adverse effects."

Understanding these terms helps distinguish the process of managing risk from the analytic process of assessing risk using the equation:

$$risk = likelihood + consequence$$

Likelihood refers to the probability that "a specific event or outcome, measured by a ratio of specific events or outcomes to the total number of possible events or outcomes." AS/NZ 4360:2004 defines *consequence* as: "The outcome of an event expressed qualitatively or quantitatively, being loss, injury, disadvantage or gain."[6] Likelihood and consequence are evaluated in the analysis phase of the risk management cycle. This analytic cycle comprises five phases shown in figure 11.4.[7] Step by step:

1. Use two tools—in the form of scales—to evaluate a target's risk rating (i.e., the asset under consideration). These two scales consist of a likelihood scale (see table 11.6) and the consequences scale (see table 11.7).

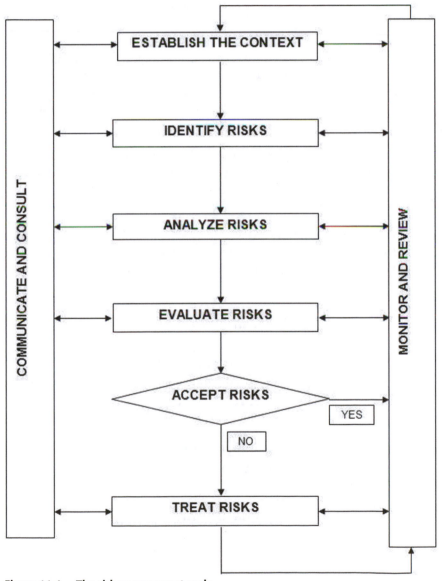

Figure 11.4.   The risk management cycle.

Table 11.6. Typical Example of a Likelihood Scale

Rank	Likelihood	Descriptors
A	Almost certain	The situation is expected to occur
B	Likely	The situation will probably occur
C	Possible	The situation should occur at some time
D	Unlikely	The situation could occur at some time
E	Rare	The situation would only occur under exceptional circumstances

2. The results of these two assessments are then fed into a risk rating matrix (see table 11.8) that returns a risk rating coefficient.
3. Finally, the analyst looks up the risk rating coefficient on the risk evaluation scale (see table 11.9) in order to determine what actions (if any) are required.

In addition to the descriptors listed in the tables, a set of conditioning statements may be necessary, along the lines of those contained in table 11.5. This also applies to the descriptors contained in table 11.7.

An example of low risk events includes:

- An event that would occur rarely and would result in insignificant consequences (reflected in table 11.8 as E1); or
- An event that is unlikely to occur and would result in minor consequences (reflected in table 11.8 as D2).

Examples of high risk situations include:

- An event that would occur rarely but would result in catastrophic consequences (reflected in table 11.8 as E5); or
- An event that is likely to occur and have minor consequences (reflected in table 11.8 as B2).

Table 11.7. Typical Example of a Consequences Scale

Rank	Consequences	Descriptors
1	Insignificant	Will result in little disruption
2	Minor	Will result in minor disruption
3	Moderate	Will cause considerable inconvenience
4	Major	Causes noticeable impact
5	Catastrophic	Causes function/services to totally fail with high impact

Table 11.8. Typical Example of a Risk Rating Matrix

Likelihood	Consequences				
	1 *Insignificant*	*2* *Minor*	*3* *Moderate*	*4* *Major*	*5* *Catastrophic*
A Almost certain	Moderate	High	Extreme	Extreme	Extreme
B Likely	Moderate	High	High	Extreme	Extreme
C Possible	Low	Moderate	High	Extreme	Extreme
D Unlikely	Low	Low	Moderate	High	Extreme
E Rare	Low	Low	Moderate	High	High

Treating Risk

Once each risk is assessed in this way, they can be positioned on the risk rating matrix (see table 11.8) so they can be compared with each other in order to prioritize treatment options. Take for instance the following events considered by troops stationed in Country Q:

- The risk posed by a person-borne suicide bomb to a public meeting place could be located at C5 (possibly with catastrophic consequences—therefore, it is an extreme risk); or
- Violence as a result of a street demonstration could be located at B3 (likely with moderate consequences—so the risk is high).

The scale provided in the risk rating table (see table 11.9) is useful for judging whether the analyst recommends accepting the risk or treating the risk (and, if so, to what extent). Without the risk assessment process, the recommendations of the analyst could be called into question as an overreaction or, equally, deemed an underestimate of the seriousness of the situation. These models curb subjectivity to some extent by providing transparency about how analysts make their calculations.

According to Emergency Management Australia (EMA), some treatment options for critical infrastructure include: awareness and vigilance, communication and consultation, engineering options, monitoring and review, resource management, security and surveillance, and community capability and self-reliance.[8]

Table 11.9.   Typical Example of a Risk Evaluation Scale

	Risk Rating and Suggested Actions for Treatment
Low risk	Manage using standard operating procedures
Moderate risk	Outline specific management actions that need to be taken
High risk	Create a business contiguity plan and a response plan (test annually)
Extreme risk	Urgent actions are necessary (in addition to those per high risk)

Although the risk rating (see table 11.9) shows what is a generally accepted distribution of risk levels,[9] analysts will need to make their own judgments as to where these transition points take place. Many times this will be a topic for discussion with the employing agency or a matter set by policy. But by using a systematic approach to risk management, analysts can reduce the likelihood and lessen consequences through the application of technology, science, or personal or collective effort.

PPRR PLANNING

There are four elements to PPRR policy development—prevention, preparation, response, and recovery. Prevention considers the risk and tries to implement ways that could stop it from happening. Preparedness acknowledges that despite preventative measures, the event may still occur, so one should prepare for it. If it does occur, response is that part of the plan that deals with how agencies will mobilize and take action (and what type of action, etc.). The final element provides guidance for recovery operation. This aspect of the plan anticipates the worst-case scenario. That is, preventative measures have failed, and preparation measures may have mitigated the impact to some degree, but it still occurred; response has contained and brought the event to an end, but it is now time to recover from the event's effects.

Even though this chapter addresses PPRR from a counterterrorist point of view, it should be borne in mind that when planning for one type of event, it is prudent to consider actions to cover what is termed *all hazards*. For instance, if an analyst is considering the impact of a particular terrorist event, then he or she should consider the event occurring as a result of nature—wildfire, flood, earthquake, and so on.

When compiling a PPRR plan, try to avoid constructing the plan in such a way that elements form either conceptual or real barriers between them—there is usually no clear delineation between the elements, though they may be expressed in these terms. Also, bear in mind that not each element will carry the same weight of importance—the four elements may not be equal. In fact, some elements may not have any strategies or treatments, or few, or minimal.

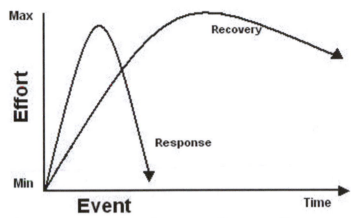

Figure 11.5. Comparison of response and recovery efforts.

Further, although the elements are cited in a sequence—PPRR—they may be put into action at the same time; for instance, response and recovery can (and should) start at the same time as they are inextricably linked (see fig. 11.5).

Finally, though the language appears to contain action-oriented terms, the treatments do not have to be physically-based options. Options involving social dimensions are also needed as, arguably, people are the ultimate targets of a terrorist attack (recall the basic philosophy of the terrorist: *kill one, frighten ten-thousand*). Analysts should try to keep their thinking about treatments broad and innovative.

KEY WORDS AND PHRASES

The key words and phrases associated with this chapter are:

- All hazards;
- Attractiveness;
- Capability;
- Coefficient;
- Consequence;
- Desire;
- Ease of attack;
- Expectation;
- Impact;
- Intent;
- Knowledge;
- Likelihood;

- PPRR;
- Resources;
- Risk;
- Threat;
- Threat agent;
- Threat communities;
- Threat profile;
- Treatment; and
- Vulnerability.

STUDY QUESTIONS

1. List the elements that comprise a threat analysis. Describe each, and explain why each is important to understanding a threat.
2. List the elements that comprise a vulnerability analysis. Describe each, and explain why each is important to understanding the concept of vulnerability.
3. List the elements that comprise a risk analysis. Describe each, and explain why each is important to the understanding of risk.
4. List the elements that comprise a PPRR plan. Describe each, and explain why each is important to counterterrorism.

LEARNING ACTIVITY

Suppose the agency in your jurisdiction that is responsible for monitoring threats of subversion and terrorism has assessed that the water pipeline that connects the city's drinking supply (say, a reservoir of some description) to your city's population is vulnerable to attack by the notional Omen Martyrs Faction. Using PPRR, devise a plan that considers each of the elements.

NOTES

1. U.S. Department of Defense, *Joint Publication 1-02, Department of Defense Dictionary of Military and Associated Terms* (Washington DC: Department of Defense, October 18, 2008), 132.

2. Carl Hammer, *Tide of Terror: America, Islamic Extremism, and the War on Terror* (Boulder, CO: Paladin Press, 2003).

3. Victoria Herrington, personal communication, May 3, 2009.

4. International Organization for Standardization, *Guidelines on Principles and Implementation of Risk Management*, ISO 31000 (Geneva, Switzerland: ISO, 2007).

5. Standards Australia, *The Australian and New Zealand Standard on Risk Management*, AS/NZS 4360:2004 (Strathfield, Australia: Standards Associations of Australia, 2004), 4.

6. Standards Australia, *The Australian and New Zealand Standard on Risk Management*, 4.

7. Standards Australia, *The Australian and New Zealand Standard on Risk Management*.

8. Emergency Management Australia, *Critical Infrastructure Emergency Risk Management and Assurance*, 2nd ed. (Canberra, Australia: Attorney General's Department, 2004), 43.

9. Queensland Government and Local Government Association, *Local Government Counter-Terrorism Risk Management Kit* (Brisbane, Australia: Queensland Government and Local Government Association, 2004), 16.

12

҉

Presenting Spatial Data

This topic focuses on three ways of presenting spatial data:

1. Maps;
2. Overlays; and
3. Mosaics.

MAPS

Basics

A map is simply a two-dimensional graphic representation of part of the earth's surface. It is drawn to a scale (except for those that are created in the field and termed *mud maps*) by cartographers and is always oriented with a bearing to north. Mapmakers incorporate certain devices into the map in order to make reading easy—universal symbols for various features in the physical world as well as colors and lines to augment the representations.

While some rare maps are three dimensional—being constructed out of sand, clay, wood, plastic, paper, or cardboard—our discussion here will limit itself to two-dimensional, paper-based maps and digital maps. These can be produced by even the most modest forms of spatial analyses, and the results can easily be displayed for decision makers' consideration.

Maps are important as they allow an analyst to simultaneously display a number of pieces of information that may prove confusing in narrative form—recall the adage *a picture is worth a thousand words*. Further, through the medium of a map, analysts can conduct their analysis-in-chief—say, comparing distances and lines of sight and evaluating access, travel routes, and so on. Because each map contains numerous data items, it is a self-contained database that shows the existence and location of many ground features as well as the distance between them.

Arguably, the most frequent user of maps in intelligence work is the military. Although some use is made by law enforcement agencies with the mapping of crimes and crime hotspots, its use is not universal. If a map is to be incorporated into an intelligence report, it is imperative that the analyst knows the cartographic skill level of the reader. If, for instance, the reader has no knowledge beyond reading a city street map, then this is the maximum level of sophistication that should be presented in the map. More detail may cause the reader to lose interest, and therefore, the analyst will lose his or her audience.

Sources

Maps can be sourced from many places—from government departments that are responsible for land surveying to commercial companies that specialize in the production of high quality maps. At the time of this writing, the Internet featured several digital sources of maps, including maps that show satellite images of the earth. The resolution of these images is good, and the user can select the level of magnification required. These satellite images can also be displayed as a simple map with terrain features or in composite form (i.e., terrain and image). Other Internet-based mapping systems available at the time of this writing included those that showed street-level maps (views) of cities around the world and a virtual view of the oceans' landscape.

Through the use of desktop computer workstations and software packages, analysts can convert paper-based maps to digital images for projection in oral briefings, incorporate them into digitally created documents, or save them to a secure intranet site within the host agency (e.g., in some form of electronic knowledge base).

Aerial Photographs

Aerial photographs are those photographs taken from any of a variety of airborne platforms, such as fixed and rotatable wing aircraft, unmanned drones, hot air and gas-filled balloons, and satellites. The aerial photograph can be used as either a supplement to a cartographic map or as a map substitute.

Why would an analyst consider displaying intelligence in the form of aerial photography? On the one hand, a topographic map may have been created some years previous and, hence, could be obsolete. On the other, an aerial photograph taken recently (via the analyst's information collection plan at the start of the research) shows any changes in landscape as they occur.

In this regard, a display of aerial photographs and maps provides more information than either alone. Photographs can be obtained readily, whereas maps take months to prepare. Changes can be noted and compared daily or even hourly if the issue is that important (e.g., the Cuban missile crisis of October 1962). The downside however is that aerial photographic interpretation is both an art and science that takes considerable skill and training.

OVERLAYS

Overlays are used to display additional information about the terrain or activities within the geographic area bound by the map. Overlays can be used for both maps and aerial photographs. An illustration of a map overlay, from the U.S. army's field manual entitled *Map Reading and Land Navigation*,[1] is shown in figure 12.1.

Figure 12.1.　Example of a military map overlay with notation in the margins.

Physical overlays are usually clear sheets of plastic-like material that allow the analyst to hand draw or annotate the map/photo with standardized symbols or codes. They are most commonly used in operational theaters that are characterized by quick decisions and frequent changes of plans. Marks on the clear sheets can be erased and redrawn as events or plans change. Several overlays can be used to "build" a picture of various activities or present different tactical options.

Anything that can be presented in a hand-drawn overlay can be presented electronically using a software package. This method is most effective in oral briefings where an image is shown, and with the click of a mouse button, addition information can be overlaid, thus emphasizing the point the analyst wants to make.

MOSAICS

A mosaic is created by combining two or more overlapping graphics in such a way that they form a single picture. Graphics can be maps, aerial photographs, and vertical photographs (including those taken at low- and high-oblique). However, the majority of mosaics are of a photographic nature as they offer operational commanders and field operatives a panorama-like view of an area under investigation.

Since their employ during World War I, mosaics have been useful in displaying spatial data and, hence, vital intelligence. With today's software packages, photographs can be combined with terrain data so that a digitized image can be constructed to produce a three-dimensional, mosaic-orthogonal map.

These mosaics are valuable in training field operatives in such techniques as border crossing and covert insertions—techniques that require the operative to enter a hostile country (or in the case of law enforcement, a violent or dangerous neighborhood in an urban area). These computer-generated, mosaic-orthogonal maps can be extensive, covering tens of thousands of square kilometers.

KEY WORDS AND PHRASES

The key words and phrases associated with this chapter are:

- Aerial photograph;
- Map;
- Mosaic;
- Mud map;
- Oblique; and
- Overlay.

STUDY QUESTIONS

1. Define the term *map*, and give three examples of common maps.
2. Describe a situation where an intelligence analyst might use an overlay.
3. List the main features of a mosaic.

LEARNING ACTIVITY

Suppose you have been asked to construct a map overlay to assist field operatives in a covert operation. Using a street map for the city in which you live as an example, use some imagination in constructing an overlay that shows the following detail: a muster point for pre-operation briefings, the location of the target, any potential hazards (countersurveillance points), and an exfiltration route. Annotate the overlay with an appropriate security classification.

NOTE

1. U.S. Department of the Army, *FM3-25.26: Map Reading and Land Navigation* (Washington DC: Department of the Army, 2001), figure 7-2.

13

✺

The Intelligence Research Report

This chapter focuses on several topics relating to intelligence reports including:

1. Writing reports;
2. Key parts of a strategic intelligence report;
3. Example of an operational intelligence report;
4. Example of an oral intelligence briefing;
5. Target profile template;
6. Tactical assessment template;
7. Correct use of intelligence terminology;
8. Use of tables and figures in reports;
9. Dissemination with regard to the concepts of "need-to-know" and "right-to-know"; and
10. Some final thoughts on writing reports.

WRITING REPORTS

The end point of the intelligence cycle is to produce a *report* of some description. Often termed an *intelligence product*, the central purpose of the report is to inform the decision maker about the problem, how the analyst approached the inquiry, what was learned, what the information means, and what implications the findings might have for the future. Reports take two forms—written reports and oral briefings. Each may contain a third form of report, which is graphics. (See table 13.1.)

187

Table 13.1.   Summary of Report Types

Products	Form
Operational and tactical reports, memos, minutes, strategic studies, and estimates	Written
Briefings	Oral
Charts, overlays, and situational maps	Graphic

Rare is the analyst who can produce a word-perfect report in one draft. Writing an intelligence report is a formative process, not summative. The analyst should not be discouraged if it takes several passes at redrafting before arriving at the final version. During the process, the analyst should seek feedback and advice from colleagues—doing so can only strengthen the document's message.

The objective should be an error-free report with no awkwardness that will distract the reader from the content. A good approach is to read the text through the eyes of the intended audience. Will they understand what is being said? Can they follow the thought sequence? Is it well laid out, and is the organization logical? The length and format of reports vary according to the agency, the type of information the audience wants to know, and the message that is trying to be conveyed (some examples and templates are provided further on in this chapter).

On the point of academic rigor, analysts should ask themselves: does the intelligence report reflect sound research and analysis? The research methods and the analytic techniques used need to be thoughtfully considered as to their appropriateness for the research question or hypothesis. Put another way: will the research and its findings withstand critical appraisal by peers? Reading widely and having a substantial personal library that contains texts on applied research methodologies are keys to being successful as a scholar spy.

STRATEGIC INTELLIGENCE REPORT TEMPLATE

Abstract

An abstract is a summary of the intelligence report that provides the reader with an overview of the research's purpose and findings. It appears at the start of the report, before the introduction or background. In the open-source literature, indexing services will often use the abstract as a way of cataloging the report for other researchers to find.

Title

The purpose of the report is usually to inform decision makers who authorized the intelligence research project (e.g., a special project), or key personnel receive such reports as a matter of course (e.g., routine briefings). Therefore, the report's title should capture the reader's attention but, at the same time, accurately reflect the essence of what's contained in the report.

Aim of the Study and its Significance

This section should contain a succinct statement of the analyst's *research question, statement of the problem,* or *statement of controlling purpose.* It should be worded along these lines: "The purpose of this research is to . . ." The statement should not encompass everything related to the issue; it should limit itself to, for example, examining the relationship between two concepts (i.e., variables)—for instance, "An increase of police on the street leads to a reduction in victim reported crime." The research question should be set in the wider context of the issue under investigation to highlight its significance, and it needs to be stated clearly, ideally in its own subsection (e.g., a subsection entitled *rationale*).

Literature Review

The literature review is an essential part of all intelligence research projects. This section serves several purposes, but the most important is that it places the issue under investigation in context. It establishes what has been discovered before, especially in what is viewed as the landmark studies pertaining to the topic and what problems those researchers encountered. It also identifies the knowledge gaps that still exist.

This section also needs to contain a discussion about the study's theoretical base and the presumed relationship between the variables. This part of the report will need to demonstrate that the analyst has an appreciation of what research has already been conducted by describing the key concepts and terms so that they can be operationalized in the report's methodology.

Methodology

Arguably, this section is the most important in the intelligence report. If the analyst has carefully crafted the research question and placed it in its theoretical framework, it will guide the research design. Such designs include:

- Evaluation (to plan intervention programs/operations);
- Case study (what is going on);
- Longitudinal study (has there been any change over time?);
- Comparison (are A and B different?);
- Cross-sectional (are A and B different at this point in time?);
- Longitudinal comparison (are A and B different over time?);
- Experimental and quasi-experimental (what effect does A have on B?); or
- A combination.

Analysts need to define the concepts they are studying so that they can be measured (i.e., operationalized). Analysts also need to identify what data they will require (whether these data will be from primary or secondary sources; whether they will be qualitative, quantitative, or both; if the data will come from open-sources, such as surveys, interviews, empirical observation, or via covert sources, etc.) and how these data will be collated and analyzed (e.g., statistically or content analysis) to test the hypothesis.

Analysts will need to turn their minds to the related issues of sample size, how they will control for confounding variables (i.e., the potential that observations are due to something other than what is being measured, and what limitations these extraneous influences might present for the research (e.g., possible alternative explanations for the relationship between A and B) or the limits inherent in the data, and so forth.

Results

Generally speaking, the results section of a strategic intelligence research report will be presented in three phases. The first is a description of how the data were prepared for analysis. This section needs to be brief and focus on the more unique aspects of the analytic technique.

This is then followed by descriptive statistics of the most relevant or important information. But again, these need to be brief to avoid overwhelming the reader with volumes of results—doing so will run the risk of confusing them by inadvertently obscuring the report's central line of reasoning (i.e., the idiom "miss the forest for the trees" applies in this situation). These descriptive data need to be carefully considered and well organized by such techniques as summary tables and graphs, maps, or other diagrams.

The third phase is linking any inferential analyses to the research question or hypothesis. If statistical analyses were not used, then this section would discuss the results of any of the specialized analytic techniques that were employed to test the hypothesis, such as SWOT, PESTO, force field, competing hypothesis, or other analytic techniques.

Ethical Considerations

In the social and behavioral sciences, if a study involves human subjects, then approval to conduct research has to be granted by an ethics committee. But as we are discussing secret intelligence research—not research that will be disseminated via public avenues (e.g., scholarly journals)—the issue of ethics takes on a different dimension.

As such, questions like, "Will there be confidentiality issues regarding the data or the findings the study will produce?" need to be considered. Could the study impact national security? Could revealing the findings jeopardize ongoing operations (or collection techniques) or the identity of key informants/witnesses/undercover operatives?

If it is not secret intelligence research but some form of strategic planning, then there may be questions about issues of privacy, intrusiveness, and appropriateness (e.g., physical or mental distress) that could arise. Analysts don't have to answer each of these questions in this section, but nevertheless, they need to consider the issue of ethics holistically.

Conclusion

Put the study's conclusions into a few paragraphs or a page or two, depending on the length of the report. This section is intended to reinforce why the research was important and what the ramifications might mean for field operations or strategic policy.

References Cited

List the references that have been used in compiling the research report using the Harvard style of referencing. Once complete, check the document for consistency in referencing, and double-check the report for inadvertent errors, such as not citing other scholars' intellectual work.

EXAMPLE OF AN OPERATIONAL INTELLIGENCE REPORT

SALUTE is a U.S. military acronym for size, activity, location, unit/ uniform, time, and equipment. It is a type of operational report about captured or foreign material and is prepared by the unit that acquired these goods. The report can be presented in either an oral or a written format. Because this type of report is filed by a field operative, its only analysis is in the form of descriptive data that accord to the letters that make up the acronym. An example of a SALUTE report taken from the U.S. army's field manual entitled *TECHINT: Multi-Service Tactics, Techniques, and Procedures for Technical Intelligence Operations*[1] appears in the following:

To: G-2 V Corps DTG: 230900Z Aug 89
From: G-2 24th Inf Div (Mech) Report No: 07-0623

1. Size: Not applicable.
2. Activity: Capture of shoulder-fired laser target designator by 1/64th Armor Bn 2d Bde 24th Inf Div (Mech) (include capturing unit).
3. Location: Town of Al-Dahran (UTM EH556937) (as a minimum always give grid coordinates).
4. Unit: 3d Republican Guards Regiment (include enemy unit if known).
5. TIME: Item captured on 230230Z AUG 98 (always use ZULU time).
6. Equipment: One laser target designator and sighting device (give best possible description).

EXAMPLE OF AN ORAL OPERATIONAL INTELLIGENCE BRIEFING

SMEAC (pronounced *smee ack*) is an acronym for situation, mission, execution, administration and logistics, command and communications (or signals) used by police and emergency services. Sometimes variations of this acronym exist, depending on the agency involved.

It is usually delivered in an oral intelligence briefing for advice to the troops who are preparing to conduct an operation. Although a SMEAC is typically delivered in this manner, it can also be delivered in a written report format for those higher up the command structure to advise them of what is happening in the operational environment. Note that it is like a SALUTE report in that it is factual and contains no analysis, though it is a product of an analytic process. Because it is an operational intelligence report, the facts of the matter are most important, and the analysis that feeds the preparation of the report is the responsibility of the analysts who issue the SMEAC.

Situation: A summary of the current situation.

Mission: A single sentence statement about the task's objective or what needs to be done.

Execution: The basic plan to accomplish the mission. Although presented in simple terms, it needs to contain as much detail as possible so that troops carrying out the mission can do it successfully. This aspect of the SMEAC may contain visual aids such as diagrams, maps, and overlays.

Administration and logistics: Composition of the teams that will be involved, how the plans will be executed, what the arrangements are for food, water, fuel, sleeping, equipment as well as the paperwork for logs and other record-keeping requirements.

Command and signals: A clear summary of who is in command and the legal authority for carrying out the mission. Directions need to be given as to how and when teams will communicate (e.g., the two-way radio channels or frequencies, mobile telephone numbers, code words to be used for predefined situations, etc.).

TARGET PROFILE

A target profile is an intelligence report that summarizes information about a specific target. The target can be an individual, but it could also be a company or organization. In the case of the latter, the report may be entitled a "criminal business profile," "terrorist organization profile," or similar name. A variation to the target profile is the *problem profile*—this is a report that focuses on an issue, not a person or organization. An example could be a series of crimes occurring in a geographic "hotspot." In any case, the target can be either the subject of a current inquiry or an emerging target for a proposed investigation. The report summarizes what is known about the target and, in doing so, identifies information gaps, which feed into a collection plan for additional data.

Target profiles often provide field operatives with a range of options regarding possible intents (i.e., hypothesis derived from inductive reasoning) or possible ramifications if the target continues the activity at the central of the report's concerns (e.g., risk assessment). In doing so, a good report will prioritize the need to make further inquiries about the target in ranked order with other targets under investigation so that intelligence unit managers can allocate resources.

A target profile consists of several sections that are arranged to tell a story: background, personal details, criminal record (in the case of a law enforcement target assessment), physical environment, analysis, and target planning. There may also be attachments appended to the end of the report.

Context is the background to a problem or situation. It allows analysts to make sense of why an event is happening by allowing them to explain key contributing factors.

Background

The background section of the profile contains a short statement of the report's aims or objectives, scope, and a brief description of how the target fits into the broader picture—the *context* (e.g., motorcycle gangs in the region). In addition, this section provides a short statement about the legal basis or agency policy, which gives the analyst authority to compile the report, and the name of the authorizing officer (e.g., the intelligence unit's manager).

Personal Details

This section provides a descriptive analysis of the target (sometimes referred to as the *person of interest* or abbreviated as POI). It will contain a biography and physical description of the target and, if available, a photograph. Any aliases used by the target are included along with a description of the target's usual or last practiced occupation, business address, business affiliations, and so on. Other pertinent facts, such as driving licenses, trade licenses, and vehicles owned or regularly driven (in the case that they do not own a vehicle), appear in this section.

The personal details section can also contain details about the target's social and/or psychological functioning. This could take into account information about places the target frequents, personal friends, close family connections, and business associates.

Business Details

A range of information about the target's own business or the businesses they are associated with appears here. Rather than being prescriptive about what should be included in this section, suffice to say that any detail that bears directly on portraying any of the five WHs—who, what, where, when, why, and how—associated with the aim or goal of the report should be included. By way of example, in a money laundering case, details should be included on off-shore banks, electronic fund transfer arrangements, businesses and nonprofit organizations in the cash flow chain, key personnel, financial information, and company data from the government agency responsible for registering companies in the jurisdiction, and so on.

Criminal Record

In addition to outlining the target's criminal history, this section could contain a list of court appearances, how they responded to previous probation or parole orders, and, if currently on bail, bail conditions and reporting arrangements. There may be intelligence from a prison or jail intelligence unit that could prove insightful—if available, consider including it.

If the target has prior convictions or has been arrested for violence or firearms offenses, these details need to be highlighted as an occupational safety issue for fellow officers who might have to deal one-on-one with the target. Further, the types of crimes or the frequency of offense may be relevant as to why the target profile is being developed, and therefore, it too should be a feature. If agency records note the target's modus operandi, then this information will be another important aspect for inclusion.

Physical Environment

Observations made by surveillance operatives appear in this section of the profile. Details about the target's physical space complements what is known about his or her social and psychological makeup (e.g., arrests and convictions can be considered external manifestations of their psyche). Moreover, surveillance operatives may have noted some of the ways the target practices countersurveillance or has put in place other intelligence countermeasures. Surveillance reports may have also commented on whether these precautions are conducted regularly, on an ad hoc basis, and whether they appear to be effective.

Analysis

This is the "engine room" of the report. The material that appears in the sections just discussed is descriptive in nature. However, in this section, the analyst applies one or more analytic tools to these data to derive some insight. It is sometimes described as the "so what" section. That is, what do all the previous facts mean, and what implications do they have? Depending on the issue or allegation, it might be as simple as positioning the target within the wider criminogenic landscape, or it might be as sophisticated as detailing where the target fits within a transnational crime syndicate.

This is done via analysis; analysts cannot rely on "gut feel" or base it on "experience" or "belief." If a conclusion is drawn, then analysts must be able to demonstrate how they arrived at this finding. For instance, if an assessment centers on the risk the target poses to the community (or field operatives, etc.), then a risk analysis must be conducted. How elaborate this analysis is depends on the facts and the issues surrounding the target as well as the aim of the report. But generally, for the purposes of a target profile, it can be a simple, straightforward table examining the likelihood and consequences of key hazards (i.e., the sources of risk).

Even if the data are "thin," at a minimum an analyst can conduct a SWOT analysis that examines the strengths, weaknesses, opportunities, and threats posed by the target. Such an analysis can highlight missing data and form the basis for recommendations for additional data collection (via a new information collection plan). If business and financial information is presented in the business section of the report, the analyst could then present some form of financial analysis or network analysis of the key personnel involved.

The analyst can insert additional analyses depending upon what data were described in the preceding sections, but one important analysis that the analyst should always include is that of countermeasures (if any). A simple force field analysis of the precautions taken by the target when using landline and mobile telephony, voice over internet (VOiP) communications, data transmission (including email), two-way radio, and satellite telephony would be helpful for planning how field operatives are tasked in the future. If these are missing, it needs to be highlighted. Include also observations made when the target makes face-to-face contact with "significant others" (does the target practice countersurveillance techniques in these situations?). Are codes, ciphers, or encryption used when they communicate?

Target Planning

Having presented the known facts and analyzed these to gain an understanding into the target's activities, this section takes the insights developed and answers the question of "where to from here?" or "so what do we do now?" The planning section usually starts off with a short narrative summarizing the findings in the context of the overall aim and then discusses practical and realistic ways of addressing the problem.

This is usually done by providing a range of policy or operational options—from the "do nothing" option to an option that could be described as a "Rolls-Royce standard." Unless, there are large political pressures being applied because of the actions of the target, the Rolls-Royce option is normally outside most agencies' budgets, so one of the middle options is going to be more attractive (acknowledging that there will be some trade-offs in effectiveness and/or efficiency which should be noted in the report). The bottom line is to present decision makers with a way of stopping, disrupting, or reducing the illegal, harmful, or otherwise detrimental effects of the target's activities.

The do-nothing option, although at first glance appears an option that would be dismissed outright, could be a viable choice where further time is required to either gather critical pieces of information or where all necessary data are in hand and arrest, capture, seizure is the next and final step. If more information is required, then the analyst must be mindful of the costs that will be incurred in gathering these data and the likelihood that such intelligence gathering will benefit the analysis and improve the recommendations of the report in a material way.

Attachments

Because a target profile is usually a short, sharp, and focused report, attachments should be kept to a minimum. A few examples of what might appear in an appendix are: a map, a photograph, or an organizational chart showing complex relationships that would be confusing if they appeared in a narrative form in the body of the text.

TACTICAL ASSESSMENT

A tactical assessment is a type of intelligence report that takes a wider view of a situation than does the target profile. Whereas the target profile focuses on a particular individual (or company or organization), the tactical assessment looks at the problem in a broader way. The audience for a tactical assessment would be a whole-of-agency group that deals with tasking or an inter-agency group responsible for coordinating assets across several organizations (which could include agencies at different levels of government, or sectors—e.g., military and law enforcement).

For example, if a target profile examines Mack DaKnife who is involved in drug trafficking, then a tactical assessment may not only look at where DaKnife fits into this picture but the extent to which the drugs are trafficked, those involved (buying and selling), the social and economic impact on the community of the direct effects of the illicit drugs, and the indirect and consequential effects of the ill-gotten profits on, say, corrupting legitimate business, political, or regulatory officials.

Although this description of a tactical assessment has the hallmarks of a strategic assessment, it falls short of its companion because it focuses on short-term objectives that would result from immediate action to prevent further illegal or otherwise unwanted activity. A strategic assessment looks at a longer timeframe and usually features several recommendations that in combination need to be put in place in order to defend against the threat or treat the risk/hazard.

Agencies will have their own house style for this type of report (and it may even be known under a variation of this name), but it will follow this template to a large degree. Analysts may optimize the sections and format to suit the intelligence project they are working on. Because the life-span of a tactical assessment is short, a new or revised report may have to be produced weekly or every two weeks.

Introduction

The report's introduction section will contain a statement of the report's aim or objective as well as a description of how the problem or issue under investigation arose. Like a target profile, this section needs to provide an acknowledgment of the legal basis or agency policy that provides the authority for compiling the report and the name of the authorizing officer (e.g., the chair of the tasking group or the officer in charge of the inter-agency committee).

Current Situation

The current situation section of a tactical assessment is comparable to the various descriptive sections that appear after the background section but before the analysis section of a target profile. In fact, this section may comprise several subsections of descriptive data about the problem. These data can come from other intelligence reports (e.g., target profiles), open sources (e.g., Internet), academic studies (PhD dissertations and master's theses), articles in scholarly journals, and government publications (e.g., bureau of statistics), to cite a few.

In addition to describing the phenomenon, the analyst can discuss the ramifications of the problem in its social, economic, or political context and its extent in the jurisdiction (e.g., the region, the nation, or around the globe). The analyst could discuss any progress being made, pitfalls encountered during the investigation, or the implementation of interventions to date. This information sets the scene for the next section, which is the analysis.

Analysis

As in the analytic section of the target profile discussed above, this is where analysts present the results of their analysis using techniques such as statistical analysis, network analysis, force field analysis, SWOT, PESTO, or competing hypotheses—or possibly others, depending on the scope of the research question.

Prognosis

The prognosis section is essentially a discussion of the analysis but extrapolates from the findings to assess what is revealed about the activity under investigation and what can be done to provide relief. The analyst may consider a change in focus from what is or was being done to a modified or totally new approach or a shift in priorities using the same interventions, and so forth.

It is common to present this discussion within the frame of a results analysis. Although termed *results analysis*, this is an exercise in deductive reasoning but in a narrative form—"thinking out loud." In order to produce a range of options for decision makers to consider, this section talks about the agency's strategic mission/goals or key performance indicators (KPIs) and how possible interventions may impact on these benchmarks. Decisions makers around the conference table can then argue priorities and resources according to "what works," "best value," or "best practice." The format of the results analysis can follow the five I model:

Intelligence. Information gathering, collation, and analyzing;

Intervention. Tactics to block, disrupt, weaken, or eliminate "the problem";

Implementation. Translating the goal of the proposed intervention (theory or principles) into practical methods on the ground;

Involvement. Ways to get other agencies (or companies, organizations, and individuals) to contribute somehow to being part of the implementation of the intervention(s); and

Impact. How the problem will be evaluated and by whom. The evaluation may be simple or complex, but as the problem is one of a tactical nature, a basic evaluation is most likely all that is needed (i.e., an output-based evaluation rather than one that is outcome focused).

Recommendations

Intelligence managers dealing with tactical issues that are within the scope of this type of report will require options for consideration. Stemming from the previous section, the analyst needs only restate the range of options available. This can appear in the form of a bulleted list to simplify what is possible. If there is a preferred option, this can be highlighted in some way—for instance, appearing first in the list with the other options appearing in a list below in diminishing order, with the least preferred at the end.

Appendixes

Because a tactical assessment is another type of focused report, the attachments that may be included need to be kept to a minimum. Examples are similar to a target profile—a map, a photograph, or an organizational chart showing complex relationships that can be more helpful than a drawn-out narrative.

General Considerations for Report Writing

When writing the intelligence report, avoid placing facts, inferences, recommendations, and analysis within the same sections. Consider using an "hourglass" approach to constructing the report. That is, start from the general, work to the specific, and then end back at the general—like the shape of an hourglass.

Each section needs to help explain the "story" so the narrative flows logically. If there is a section of the report that discusses the current situation, it could contain the facts that are known. The analysis section that follows takes these facts and subjects them to one or more analytic techniques so that insight can be developed. This analysis should also place some level of likelihood/probability to these scenarios (or hypotheses) with discussion about the limits (e.g., based on, say, a risk or inferential analysis). This narrative would then lead the reader to the plan/recommendations section—like an hourglass. Or simply, it's a story that has a beginning, middle, and an end.

Unlike fictional stories, the beginning of an intelligence story contains facts—who, what, where, when, why, and how (five Ws and one H). The middle tends to contain analysis, and the end contains the report's conclusions, recommendations for action, or policy options, all of which can appear in a variety of forms (often prescribed by the employing agency in a standardized template).

Analysts should resist the temptation to append their analytic results at the end of the report as it could mean they failed to refer to them in the narrative (i.e., to help explain the story). Alternatively, referring the reader to peruse the appendices is getting the reader to do the job of the analyst.

In the same vein, tables of data or matrices should not be simply dropped into the analysis section of the report as there is often a lot of detail contained in these. It is better to summarize the key aspects of these analyses, and in case the reader wants to see the big picture or how you arrived at your conclusion, place the diagram or other figure/table in an appendix.

The reason for suggesting this writing approach is that the key points become the basis on which the analyst will make recommendations. Presenting the study's recommendations as a set of options is an important aspect of the report. Couching the options in terms of a threat assessment, a risk assessment, a budget, or the human resources available are excellent approaches. To the reader, such features make the recommendations actionable.

These guidelines are not hard-and-fast, but following them will increase the likelihood that the message contained in an intelligence report will be understood by decision makers and, if so, more likely to be acted upon.

TERMINOLOGY

In practice, some terminology used in the intelligence arena is applied inappropriately. As practitioners, it is important that we use key terms correctly; otherwise, our reports will lack credibility. Following are a few words that typically cause difficulties. Next to each term appears a definition. No doubt there are other definitions, but what is important in considering these is that analysts focus their thoughts on writing precisely.

Intelligence. "Insight" expressed in the form of a product (e.g., report, target profile, target assessment, estimate) and the process that produces such a product.

Investigation. The assembling (i.e., pre- or during) or reconstruction (i.e., post-) of facts surrounding an event.

Truth. A subjective notion. For instance, it was once viewed that the earth was flat, and this was the accepted truth at the time. To have questioned this truth was heresy. In the criminal justice setting, truth is derived via a judicial process of proofing evidence in accordance with the rules of evidence, criminal procedures, and precedent.

Fact. Something that can be observed or experienced through one of the five senses.

Allegation. An accusation yet to be substantiated.

Suspected. Alleged.

Proof. The legal process of introducing a fact into evidence with the intent to establish the truth of something.

Standard of proof. Beyond reasonable doubt (criminal) and on the balance of probabilities (civil).

Inference. A conclusion drawn from data that have been subject to analysis.

Conclusion. An inference.

Analysis. Systematic examination that follows some scientific, mathematical, or logical procedure or process.

Assumption. Something that is considered to be true without proof.

Postulate. Something that is considered to be true and forms the basis of a theory.

Hypothesis. A theory, an explanation for something, which is then used as the basis for examination or investigation.

Speculation. Conjecture, guesswork.

Believe/Belief. Synonymous to faith, not a recognized intelligence methodology. Suggest that the word *consider* is used. *Consideration* is based on reasoning.

Consideration. Deliberation, a process of long careful thought.

Probability. Likelihood, chance of something happening. Statistical confidence is the probability that a statement is true. If probability is not based on analysis, it is conjecture. (The use of words such as "will" conjure the idea of 100 percent certainty—avoid saying "something will" unless probability analysis confirms this.)

Evidence. Something that provides an indication that something exists or contributes to the process of proving the truth of something.

Finally, different words mean different things to different people. Adjectives that invoke some level of sensationalism should be avoided in an intelligence report. The best way to do this is to "describe" or "explain," but don't excite. For instance, what an analyst might see as "extreme" may be "insignificant" to the reader.

Avoid the use of personal, subjective, or unproved ideas and ambiguous language in the intelligence report. Statements made need to be supported by evidence.

TABLES AND FIGURES

Representing complex data in a report will often take the form of a graph, chart, or table—or a picture such as a photograph, diagram, or drawing. When doing so, the analyst needs to label this symbolic representation correctly so the reader can identify it in the body of the report.

There are just two terms used to do this—tables and figures. A *table* is where data are arranged in rows and columns. Graphs, charts, diagrams, photographs, and any other illustrations are known as *figures*. They are numbered sequentially as they appear in the text and independent of each other—that is, tables are numbered as a group, and figures are numbered as another group.

DISSEMINATION

Dissemination is the object of your project; but to whom do you intend to distribute your findings? To some degree, you should have discussed this in the section on significance, but here it needs to be spelled out—will the results be for publication in an academic journal, circulated directly to decision makers, or will it be a classified report for operational commanders who may use it to initiate a criminal investigation or to guide/focus an existing inquiry? In this regard, will you formulate any intervention strategies that will form part of your report's conclusions?

Need-to-Know and Right-to-Know

The term *need-to-know* is used by the military as well as law enforcement and government agencies to describe sensitive information that needs to be protected by limiting those who receive it. According to the need-to-know doctrine, even if one has a security clearance (e.g., secret), it does not entitle him or her to have access to any or all material classified at the secret level. Before people are given access, they need to demonstrate that they have a need-to-know. In short, this is some justification that access to the report will aid an intelligence project or operational mission. (In the post-9/11 security environment, there is also the newly introduced concept of *need-to-share*.[2])

Some consideration will need to be given to not only at what level will the analyst's report be classified but also some suggestion as to whether the report (or briefing) will be beneficial to other analysts, commanders, or other agencies that might be working on the same or related issue. The idea is to discourage those with a mere curiosity but not hinder legitimate access.

Consider this historical case as an example of need-to-know: during the planning of the June 6, 1944, D-Day invasion of Nazi-occupied Europe, thousands of military commanders were involved. However, only a few, by comparison, knew the full details of the plan. The great bulk of commanders were only privy to the details that allowed them to execute their part of the invasion.

Contrast need-to-know with *right-to-know*. The latter is where a legal precedent exists, allowing a person to have access to information. Usually this precedent is embodied in a statute but may be found in common law. Further, right-to-know may extend to a freedom of information law or a similar law. Analysts should make themselves familiar with laws in their jurisdiction and how they impact what they write and to whom it is disseminated, as unintentional public disclosure may result. Such disclosure may put at risk methods or information that could jeopardize operations.

SOME FINAL THOUGHTS ON WRITING REPORTS

Analysts will frequently consider the question of what constitutes a good intelligence report at the culmination of their inquiries. Given that there are only two ways analysts communicate with decision makers—orally and in writing—deliberation will invariably center on clarity of expression.

Decision makers are people with busy agendas, tight budgets, and substantial pressures, so an analyst's report needs to be concise and precise. If a decision maker cannot find the important information in a report, then, to a large degree, the report has failed. Remember, the purpose of intelligence is to provide *insight* so the best possible decision can be made. This is why the above examples and templates have been described in this section—they have been tested many times by many agencies and by busy managers. Some key points to remember are:

- Double check that the report's conclusions and recommendations dovetail with the original aim of the inquiry;
- Comply with the agency's in-house style or template for reports;
- Do not selectively omit data items if they do not support a "preferred" position—this is not only unethical but can lead to civil and criminal charges against individuals or the agency if a court or commission of inquiry finds the intelligence investigation did not act in good faith;
- Use a number of analytic techniques to distill the data so that the clearest picture emerges;
- When formulating judgments, make it clear what the limitations are—do not give false indications;
- Follow the agency's policy on making and couching recommendations; provide options in objective terms, and avoid "rivers of blood" prognoses that manipulate decision makers (doing so is bordering on the unethical);
- Aim to write several drafts, and run each version past a colleague who is senior in years of service. This is because he or she is likely to have experienced many of the pitfalls common in presenting reports and can steer a new analyst around the "holes in the road." It is better that someone close to you critically reviews your work than to have an executive in your agency do it with the potential consequence of getting a "black mark" against your reputation; and
- Always get someone else to proofread your work—it is more likely that he or she will note spelling and typographical errors than you as the author.

In intelligence writing, there is a need to present findings in tentative terms and to avoid the temptation of absolutism.

KEY WORDS AND PHRASES

The key words and phrases associated with this chapter are:

- Abstract;
- Context;
- Figures;
- Five I model;
- Five Ws and one H;
- Graphical presentation;
- Insight;
- Need-to-know;
- Need-to-share;
- Oral briefing;
- Person of interest;
- Product;
- Right-to-know;
- SALUTE;
- SMEAC;
- Tables;
- Tactical assessment;
- Target profile; and
- Written report.

STUDY QUESTIONS

1. Describe the "hourglass" approach to report writing and its advantages.
2. Explain the standard parts of a strategic intelligence report.
3. Explain the individual parts of a report represented by the acronym SMEAC.
4. Explain the acronym SALUTE stand for, and how is it different from a SMEAC.
5. What is the difference among a need-to-know, a right-to-know, and a need-to-share? Give an example of each.

LEARNING ACTIVITY

Suppose you are requested to present the findings of your research into an issue facing your agency. To demonstrate the skills learned in this chapter, construct an electronic slide show that will be the basis of your oral presentation. Use the hourglass approach to do this, and for the subject material use the topic of report writing.

NOTES

1. U.S. Department of the Army, *FM 2-22.401: TECHINT: Multi-Service Tactics, Techniques, and Procedures for Technical Intelligence Operations* (Fort Monroe, VA: Department of the Army, 2006), B-3.
2. *Need-to-share* involves a mindset that asks the question "who needs to know this piece of information?" or "who should I be sharing this information with?" Some in the national security arena argue that need-to-know should be replaced with need-to-share. But this may be a simplistic argument as the need-to-know doctrine still has a role to play. Nonetheless, need-to-know does not exclude the need-to-share philosophy and could easily accommodate it if it is articulated correctly in policy.

14

ٯ

Ethical Considerations in Intelligence Research

This topic introduces the notions of accountability and control in collecting data secretly. It also examines the ethical issues involved in presenting research findings by looking at applied social research and how ethics in this field of inquiry compare to secret intelligence research.

INTRODUCTION

Discussions about morality, politicalization, "guideposts," principles, codes, creeds, and values are not the expected reading material for an intelligence analyst. Indeed, such a collection of ethics-based issues is a most unlikely feature of the profession's tradecraft. Yet, intelligence professionals face the dilemma of acting ethically while at the same time engaging in what some have portrayed as an unethical business—spying.

Decision making is not easy, especially when faced with a choice between options that are unappealing or distasteful. However, in intelligence work, this can be the case.

BACKGROUND

By and large, intelligence analysts are recruited with academic backgrounds in such fields as sociology, criminology, anthropology, psychology, history, political science, and the military sciences. While obtaining their professional credentials, analysts are likely to have been indoctrinated in the philosophy that they must be ethical—for instance, to take responsibility for the mental, emotional, and physical well-being of those they research.[1]

A number of questions therefore arise, and these concern both research ethics and practice ethics:

- Should intelligence analysts be bound by the same ethical guidelines as their social and behavioral science brethren?
- If so, how does one reconcile being asked to carry out secret intelligence research where the welfare of those researched is not only removed from the fore of the analyst's considerations but also is unlikely to even feature anywhere in the research methodology?
- Likewise, how do analysts restrain their personal opinions from making their way into a formal assessment?
- How do they guard against presenting their own beliefs in the intelligence product?
- How do analysts maintain professional distance and not attempt to influence decision makers through their analytic product yet still provide advice as to options?

Then there are the questions at the opposite end of this ethical issue, exemplified by what the Australian government, under the prime minister John Howard (March 1996 to December 2007), was accused of doing with intelligence assessments on Iraq. These intelligence products were purported to have been created with qualifiers to reflect the ambiguity of the situation in Iraq prior to the 2003 invasion. However, an intelligence analyst then with the Australian Office of National Assessments, Andrew Wilkie, alleged that the former Howard government skewed these findings by taking the ambiguity out of the weapons of mass destruction issue and presented what appeared to be a clear-cut situation to the public.[2, 3] How does an analyst deal with this type of problem?

SOME ETHICAL DILEMMAS

Given the breadth and scope of these dilemmas, the question that presents itself is: what guidelines should intelligence professionals follow?

Australian Institute of Professional Intelligence Officers: Code of Ethics

- To continually strive to increase professionalism, integrity, respect and recognition of the profession.
- To ensure that duties and responsibilities are carried out with diligence while maintaining the highest degree of professionalism and avoiding all unethical practices.
- To fully comply with all applicable laws and regulations in relation to official duties.
- To protect intellectual property and confidential information with strict observance of the protocols in this regard.
- To promote and encourage compliance with these standards within the profession and work environment.[4]

As there is no clear-cut answer, the best way to deal with these issues is to read widely and discuss the dilemmas with colleagues and raise these issues at staff meetings.

It would be a good starting point for the analyst to examine the issue of ethics within the intelligence community in its widest context: "'truth' is a goal, yet deception, secrecy and morally troubling compromises are often necessary."[5] Whether it is collecting data by deception, or, say, coercing criminals into becoming informants, the effects of such actions on intelligence officers can be profound. Having to decide what is "right" can result in self-inflicted psychological damage as well as suffering the real-world consequences from actions taken (reflect on what Wilkie risked by blowing the whistle on the Iraq war and weapons of mass destruction).

As for intelligence collection and analysis, a recently declassified speech given by the then director of central intelligence, Robert Gates, in the CIA auditorium in March 1992 exemplifies how politicalization manifests itself in different ways—from deliberately distorting analysis and judgments to suit the preferred line of thinking to forcing intelligence products to conform to policymakers' preconceived views. Gates draws attention to the issue of how management can apply pressure to "define and drive certain lines of analysis and substantive view points,"[6] to alter the tone or emphasis of the product or the process that created these products, or to limit alternative viewpoints expressed within.

The points made by Gates were given new currency when they were echoed in the *Report to the President of the United States* by the Commission on the Intelligence Capabilities of the United States Regarding Weapons of Mass Destruction.[7] This report pointed out that the key intelligence briefings to White House personnel and senior executives were skewed.

The report said that these intelligence products demonstrated "attention-grabbing headlines and drumbeat of repetition, [that] left an impression of many corroborating reports when in fact there were very few sources. And in other instances, intelligence suggesting the existence of weapons programs was conveyed to senior policymakers, but later information casting doubt upon the validity of that intelligence was not. In ways both subtle and not so subtle, the daily reports seemed to be 'selling' intelligence in order to keep its customers, or at least the First Customer, interested."[8] It's interesting that Gates's insights preceded the commission's findings by some thirteen years—it highlights how short corporate memories can be for "lessons learned."

Nelson Blackstock's book *Cointelpro: The FBI's Secret War on Political Freedom* is an example of how domestic intelligence operations were used to subvert political groups in the 1960s and 1970s.[9] The counterintelligence program ran for decades, ending officially in April 1971. It was argued that the primary purpose was to harass and disrupt legitimate political activity under an alarmist guise of national security (e.g., communists, the New Left, and the Ku Klux Klan).

Putting aside the unpalatable political philosophies that these groups may have advanced, Blackstock presents an equally unpalatable tail of systemic misuse of intelligence by several U.S. government agencies that participated in the counterintelligence activities. As a case study, *Cointpro* presents a springboard for ethical discussion by students of the craft of intelligence.

ISSUES WITH INTELLIGENCE DISSEMINATION

Dissemination is the ultimate object of an intelligence research project, but what is it that the analysts intend to do with their research findings? The analysts should have discussed this, but here it needs to be spelled out. Will the results be for publication in an academic journal, circulated to government policy makers, or published in a classified report for operational commanders? Will you formulate any intervention strategies that will form part of your report's conclusions?

Nations can lose sight of their purpose and jeopardize the democratic principles they are trying so diligently to protect. This same issue has been highlighted with regard to foreign policy intelligence. For instance, Roger Hilsman, former professor at Columbia University, succinctly put the case: "Finally there remains the ultimate moral and ethical question, whether the means we use will eventually corrupt our values so as to change the nature of our society just as fundamentally as if we were conquered."[10]

Another key area is that of intelligence operations—that is, covert action—by field operatives. It is important to avoid the alarmist anti-intelligence literature that abounded some years ago and focus on literature that stimulates meaningful debate. James Barry's article "Managing Covert Action: Guidelines from Just War Theory" is outstanding in this regard.[11]

Barry argues that creating a framework based on "just war" guidelines could prove a credible basis for mounting paramilitary operations as well as those involving various forms of coercion and violence in its different manifestations. Although it goes without saying that it would be unrealistic to think such an approach on its own would eliminate the controversy surrounding covert action, nonetheless, just war theory is a credible platform for announcing the fact that a government is interested in the issue of "right" versus "wrong" and addressing it in an atmosphere of openness and transparency (and in doing so, making sure that this important policy option is still viable).

> "The most popular technique for securing information is socializing with competitors in non-business settings. Business people generally view their competitors negatively, believing that they go to much further lengths than does their own corporation in gathering competitive intelligence."[12]

Finally, there are the ethical issues in some selected parallel professions that have been associated with intelligence gathering—sociology, anthropology, and business. Darren Charters's article on the challenge of ethical competitor intelligence is worthy of note.[13] (Competitor intelligence is sometimes termed *business intelligence* so that its abbreviation—CI—is not mistaken for *counterintelligence*. CI is also used by law enforcement agencies when they refer to the protection of *critical infrastructure*.) In his article, Charters provides a "tool" for gauging whether one's actions could be deemed ethical or not. His tool—based on an evaluative process—is particularly helpful for analysts who are operating in an environment that does not have formal policies or guidelines in place.

Using the acronym CHIP, Charters constructs a four-factor process by which analysts can weigh the gravity of their proposed actions—community virtues, harm, individual as end, and personal virtues. Using a matrix approach (similar in intent to a SWOT analysis), CHIP compares ethical theories by considering the planned competitor intelligence activity from the perspectives of utilitarian, Kantian, and virtue ethics. Though one could argue CHIP is not a substitute for ethical training, the model does offer a realistic means for benchmarking ethical competitor intelligence activities. Its use would certainly encourage quality and consistency in order to avoid violating professional standards (as well as legal statutes—lest we forget Watergate).

The U.S. Economic Espionage Act of 1996 criminalizes the unauthorized appropriation of trade secrets:

a) To benefit a foreign government, foreign instrumentality, or foreign agent; or
b) For financial or commercial benefit.[14]

Overall, it would be advantageous for analysts to examine the literature on the ethics of intelligence. It might help prompt the formulation of ethical codes for both research analysts and field operatives. Such codes could pave the way for analysts who need to work through the issue of "doing the thing right" when confronted with the dilemma of "doing the right thing."

But, in reality, the distinction between these two paths will no doubt be made with some temperance. Analysts should always be conscious that the final judgment will be grounded in what the people and their constitutionally elected representatives consider "necessary and proper" in the context of the threat (thinking particularly of international terrorism and transnational crimes, such as the trafficking in arms, drugs, and humans).

Suggestions put forth by the International Association of Chiefs of Police (IACP) to guard against allegations of unethical conduct by law enforcement field operatives are worthy of note. The IACP advises that in order to protect the rights of individuals genuinely not involved in criminal acts, a set of rules and standards should be adopted to deal with the recording and purging of intelligence files.[15] In the main, these guidelines urge that when an individual's association is not criminal in nature, or criminally related, the information should not be recorded. Likewise, in the case of organizations, unless an organization's "ideology advocates criminal conduct and its members have planned, threatened, attempted or performed such criminal conduct," it is both unnecessary and wrong to gather such information.[16]

Moreover, information about an individual's "sexual, political or religious activities, beliefs or opinions, or any dimension of private lifestyle" should not be collected or recorded in intelligence files unless that information is material to a criminal investigation.[17] In addition, all information collected should be evaluated prior to collation to determine its accuracy. The source also requires evaluation to determine its level of reliability. Information that is not verifiable should not be stored in criminal intelligence files, and likewise, the collection of such information should be strictly limited.

A set of standards is necessary to maintain a successful intelligence unit filing system. The elements of such a system include:

1. Specific guidelines should be established for determining:
   a) the kind of information that should be kept in intelligence files;
   b) the method of reviewing the material as to its usefulness and relevance; and
   c) the method of disposing of material purged from intelligence files considered to be no longer useful or relevant.
2. Systematic flow of pertinent and reliable information;
3. Uniform procedures for evaluating and validating information;
4. A system for proper analysis of information;
5. A system capable of rapid and efficient retrieval of all information;
6. Explicit guidelines for disseminating information from the files; and
7. Security procedures.[18]

Systematic purging of files according to such guidelines should ensure that information being collected (as per an intelligence collection plan) is related to approved projects and necessary to meet the decision maker's intelligence requirements.

Situations where data have become irrelevant because of age should not be allowed to develop. For instance, a Rand Corporation study once found an American police department retained intelligence files that "contained information on suspected Nazis, the concern of the 1940s; Communist Party membership rosters, the concern of the 1950s; Black militants, right-wing extremists, and anti-war demonstrators, the concern of the 1960s."[19]

Intelligence managers in the twenty-first century would be hard pressed to justify the retention of these types of data. Analysts should always be mindful of their targets. No matter how unpalatable the cause, they should be conscious of the difference between legal, lawful protest, and subversion and criminal activity—distinctions not always correctly made in the past.

Society of Competitive Intelligence Professionals: Code of Ethics

- To continually strive to increase the recognition and respect of the profession.
- To comply with all applicable laws, domestic and international.
- To accurately disclose all relevant information, including one's identity and organization, prior to all interviews.
- To avoid conflicts of interest in fulfilling one's duties.
- To provide honest and realistic recommendations and conclusions in the execution of one's duties.
- To promote this code of ethics within one's company, with third-party contractors and within the entire profession.
- To faithfully adhere to and abide by one's company policies, objectives and guidelines.[20]

For the most part, secrets in the context of this book are secrets of the state (or a private corporation), and as such, keeping them secret will most likely be an obligation of a legal statute or a legally binding agreement. Because of this, and unless disclosure is covered by a countervailing legal instrument (e.g., a whistleblower-type act, an order of a court, a police warrant, a direction given by a judge or magistrate), then disclosure should not be made. However, if the secret is the cornerstone of a cover-up of an illegal activity or it relates to the planning or commission of an illegal activity (or covered under a whistleblower-type law), then revealing the secret is not likely to fall into this category—in fact, there may be a legal obligation to disclose these facts to a law enforcement agency.

By way of example, take the U.S. army's field manual *Open Source Intelligence* which lists a number of areas that prohibit the carrying out of intelligence activities, specifically mentioning activities that may be a violation in law. The prohibited intelligence activities identified are the improper collection, retention, or dissemination of U.S. person information:

- Gathering information about U.S. domestic groups not connected with a foreign power or international terrorism.
- Producing and disseminating intelligence threat assessments containing U.S. person information without a clear explanation of the intelligence purpose for which the information was collected.
- Incorporating U.S. person criminal information into an intelligence product without determining if identifying the person is appropriate.
- Storing operations and command traffic about U.S. persons in intelligence files merely because the information was transmitted on a classified system.

- Collecting U.S. person information from open sources without a logical connection to the unit's mission or correlation to a validated collection requirement.
- Identifying a U.S. person by name in an intelligence information report without a requirement to do so.
- Including the identity of a U.S. person in a contact report when that person is not directly involved with the operation.[21]

KEY WORDS AND PHRASES

The key words and phrases associated with this chapter are:

- Cover-up;
- Democratic principles;
- Ethics;
- Legal obligations;
- Moral dilemmas;
- Politicalization;
- Purging files; and
- Whistle-blower.

STUDY QUESTIONS

1. In your view, should intelligence analysts be bound by the same ethical guidelines as colleagues in the social and behavioral sciences? Discuss the issues.
2. As an intelligence analyst, cite several practical ways you could exercise restraint when it comes to expressing your personal opinions in a formal assessment.
3. Discuss how analysts might maintain professional distance and not attempt to influence decision makers through their analytic product while still providing advice as to policy options.

LEARNING ACTIVITY

Suppose you are asked to carry out a classified intelligence research project. Discuss how you would reconcile the fact that you are being tasked to carry out secret research where the welfare of those being researched is not a feature in the project's methodology. Consider the legal and ethics issues, and weigh-up the importance of these factors in your overall judgment. Are there instances where such considerations are a non-issue?

NOTES

1. Max Futrell and Cliff Roberson, *An Introduction to Criminal Justice Research* (Springfield, IL: Charles C Thomas, 1988), 201–4.

2. Veteran Intelligence Professionals for Sanity and Andrew Wilkie, "Memorandum: One Person Can Make a Difference," in *The Ethics of Spying*, ed. Jan Goldman (Lanham, MD: Scarecrow Press, 2006), 188.

3. Andrew Wilkie, *Axis of Deceit* (Melbourne, Australia: Pan Macmillan Australia, 2004).

4. Australian Institute of Professional Intelligence Officers, Inc., *Intelligence Officer Code of Ethics*, http://www.aipio.asn.au/files/file/CODE_ETHICS.pdf (accessed April 25, 2009).

5. Jan Goldman, ed., *The Ethics of Spying* (Lanham, MD: Scarecrow Press, 2006), x.

6. Robert M. Gates, "Guarding against Politicization: A Message to Analysts," in *The Ethics of Spying*, ed. Jan Goldman (Lanham, MD: Scarecrow Press, 2006), 172.

7. Commission on the Intelligence Capabilities of the United States Regarding Weapons of Mass Destruction, *Report to the President of the United States* (Washington DC: U.S. Independent Agencies and Commissions, March 31, 2005).

8. Commission on the Intelligence Capabilities of the United States Regarding Weapons of Mass Destruction, *Report to the President of the United States*, 14.

9. Nelson Blackstock, *Cointelpro: The FBI's Secret War on Political Freedom* (New York: Vintage Books, 1976).

10. Roger Hilsman, "On Intelligence," *Armed Forces and Society* 8, no. 1 (Fall 1981): 129–43.

11. James A. Barry, "Managing Covert Action: Guidelines from Just War Theory" in *The Ethics of Spying*, ed. Jan Goldman (Lanham, MD: Scarecrow Press, 2006), 248–65.

12. William Cohen and Helena Czepiec, "The Role of Ethics in Gathering Corporate Intelligence," *Journal of Business Ethics* 7, no. 3 (1988): 199.

13. Darren Charters, "Business: The Challenge of Completely Ethical Competetiveness Intelligence and the 'CHIP' Model," in *The Ethics of Spying*, ed. Jan Goldman (Lanham, MD: Scarecrow Press, 2006), 362–77.

14. U.S. Congress, *Title 18 of the United States Code*, Part I, Chapter 90, 1831–39.

15. International Association of Chiefs of Police, *Law Enforcement Policy on the Management of Criminal Intelligence* (Gaithersburg, MD: International Association of Chiefs of Police, 1985), 8–9.

16. Los Angeles Police Department, *Standards and Procedures for the Anti-Terrorist Division* (Los Angeles: Los Angeles Police Department, 1984), 2.

17. Los Angeles Police Department, *Standards and Procedures*, 2.

18. International Association of Chiefs of Police, *Law Enforcement Policy*, 9.

19. Brian Michael Jenkins, Sorrel Wildhorn, and Marvin Lavin, *Intelligence Constraints of the 1970s and Domestic Terrorism: Executive Summary* (Santa Monica, CA: Rand, 1982), 5.

20. Society of Competitive Intelligence Professionals, *SCIP Code of Ethics for CI Professional*, http://www.scip.org/About/content.cfm?ItemNumber=578&navIte mNumber=504 (accessed April 25, 2009).

21. U.S. Department of the Army, *FMI 2-22.9: Open Source Intelligence* (Fort Huachuca, AZ: Department of the Army, 2006).

15

🌀

Document and Personnel Security

The topics discussed in this chapter focus on data security and associated issues for those who handle it, including:

1. Information security;
2. Handling sensitive information;
3. Security clearances; and
4. Storing, protecting, and disposing of intelligence data.

INFORMATION SECURITY

The term *information* reflects all forms of data—ideas, concepts, and plans. *Information security* is therefore concerned with the arrangements for protecting these data once they are recorded. Information security should not be confused with the same term used by the information technology industry, which has a narrow application—that is, the security of data processing hardware and software, and access to the same.

The role information security plays in an intelligence unit is to protect data and reports from *unauthorized* personnel. Keep in mind that the information held by an intelligence unit and the finished intelligence it produces are destined for a very specific client group. These customers have either a need-to-know or a right-to-know. The delicate subject matter in a unit's files makes it an attractive target to organized criminals, terrorists, corrupt public officials, investigative journalists, and a wide range of antiestablishment ideologues.

In determining the security needs of an intelligence section, we are reminded that it is not absolute security that is sought but a level of security in line with the sensitivity level of the information being guarded. However, having said that, the requirements of security should not eclipse the basic objectives of the unit and impinge upon its operational effectiveness. Security should not inhibit the flow of information or hamper the dissemination of intelligence to authorized users (i.e., recall the need-to-share doctrine).

In addressing the issue of security, there are three areas to be examined: 1) personnel, 2) physical, and 3) information. In the defense industry, for instance, standards have long been established to ensure confidential matters are not prematurely disclosed. By way of example, if the Australian navy were to develop a new electronic guidance system for its offensive ship-to-ship missiles, such a project would be expected to necessitate dozens of civilian contractors, numerous military units, and various political leaders.

The logistics of containing the myriad details associated with such a project could be a nightmare if a comprehensive set of classifications and supporting guidelines were not developed. For example, guidelines can be found in the open-source literature such as the U.S. Department of Defense's *National Industrial Security Program Operating Manual*[1] and *A Guide to Marking Classified Documents*.[2]

Similarly, in the law enforcement or regulatory context, a set of information classifications and security procedures is mandatory. An example of this is the former Australian Bureau of Criminal Intelligence's guidelines entitled *Document Security*.[3]

HANDLING SENSITIVE INFORMATION

Classification of Data

Initial Considerations

The considerations outlined in this chapter are not intended to be rigid in their adaptation but rather flexible in their approach. The reader must consider many factors before implementing the following countermeasures. Important issues that affect the establishment (or upgrading) of a security program include, but are by no means limited to, financial constraints and the willingness of staff to follow proposed procedures once enacted. There is little sense, for instance, in spending large sums of money on a state-of-the-art, monitored intruder detection system if the costs push a budding private intelligence business to the brink of insolvency. Likewise, staff may be tempted to bypass security procedures if those procedures are viewed as overly complicated or time consuming.

The security treatments discussed here can be adapted either in whole, or in part, depending on circumstances. The important fact is that the principles of security are observed and that periodic inspections are carried out to check on the standard of security practiced.

Identifying Levels of Threat

Threat identification is the initial step for any intelligence unit when establishing a security program. Likewise, it becomes an ongoing consideration once a security program has been developed. Following are three broadband sources of threat. This list is intended to give the reader an appreciation of the hierarchy of threats that an intelligence unit may face in the course of conducting "business."

Steps taken to thwart intelligence collection at what could be termed a level II threat would be sufficient to guard against any attempt by the inferior level III threat but not the reverse. This is an important factor to remember. It is critical for intelligence units to determine where their threats lie before deciding on the range and depth of security measures they will require. Furthermore, an intelligence unit's threat level may change from time-to-time due to the dynamics of its operations. Therefore, its security needs will also be required to either escalate or abate in response to these changing conditions.

Level I Threat. Surveillance by a foreign government's security or intelligence agency, or surveillance by one's own national law enforcement/ intelligence organization(s).

Level II Threat. Surveillance by a regional/state law enforcement or intelligence unit, an organized criminal group, a foreign or domestic business competitor employing a "spy-for-hire," a private detective acting on behalf of a party interested in the analyst's unit, or other professional fact finders (e.g., an investigative journalist).

Level III Threat. Nonprofessional surveillance by, say, an employee, a business associate/competitor, or another interested individual or group acting on for profit or revenge.

Classification of Information

Business and Private Intelligence

In order to foil possible attempts by a hostile agent to obtain information in the private or business sector, information about the unit's activities should be divided into classifications of sensitivity. Moreover, these classifications should be used as a guide when releasing or disseminating information. (Classification of information in the law enforcement, military, and national security sector is discussed further on in this chapter.) Information in this context means knowledge which requires protection from disclosure in addition to the protection afforded to the intelligence it produces or that is being produced. This includes such areas as:

- Research plans;
- Research methods;
- Research schedules;
- Reporting dates and timeframes;
- Strategic plans;
- New intelligence targets and research projects;
- Customer/client distribution lists;
- Details of alliances with other intelligence units;
- Policy directives;
- Operating budgets;
- Information sources;
- Information, technology, and communications; and
- Personnel (their numbers, positions, salary packages, and expertise).

The lowest classification of information, designated as grade IV, consists of information of a general and unrestricted nature. The type of information provided in company prospectuses is a good example of this. Such information would be suitable for all general inquiries and posting to a website.

The next highest classification, grade III, consists of information which should be available to clients only upon request. Information of this type is best described as information and/or material that, if disclosed to an adversary, could reasonably be expected to cause some degree of "damage" to the intelligence unit.

Moving up the scale again is grade II information. This information should be available to a unit's most important customers. Information with this designation would be information and/or material that, if disclosed inappropriately, could reasonably be expected to cause "serious damage" to the intelligence unit.

Finally, the most sensitive information, grade I, should be available only to staff with a need-to-know and government departments that have appropriate authority. Information of this type, if disclosed to an adversary, would reasonably be expected to cause "exceptionally grave damage" to the intelligence unit. (In a business setting, this classification might mean that the company's bottom line could suffer an impact of five percentage points or more.) Staff authorized to access documentation of this grade should be required to sign a "chain of custody record" in order to assure control over its content. The chain of custody record also facilitates withdrawal and destruction when the documentation is no longer required.

In order to inform staff members of a particular document's degree of sensitivity, each document should be identified with a marking indicating its grade (the Roman numerals I to IV in red ink). When marking a document, keep in mind that perhaps not the entire document needs to be classified at a particular sensitivity level. Take for instance a report compiled on a recent research project. It is considered to be ideal for public release in a future counterterrorist awareness campaign in the media (grade IV), but a page (or even several paragraphs) contains technical data about the research that is best kept reserved. That section can carry a grade II stamp, while the remainder of the text displays the general grade IV classification.

By using an information classification system, inappropriate disclosure is less likely to occur, and as the information contained in various documents becomes dated and less sensitive with the passage of time, reclassifying the information's grade downwards can then take place.[4] The extent to which an intelligence unit goes to enact a classification system is determined by its size and the overarching authority imposed on it by its mandated creator (i.e., government and military intelligence units will have standards imposed by law or regulation, whereas private intelligence units will be guided by policy). In the case of a sole intelligence practitioner or a small research firm, there will be far less need for formal arrangements when compared to large organizations.

SECURITY CLEARANCES

Screening Personnel

Arguably, all agencies hold some form of information that would be valuable to a competitor or an adversary, especially intelligence agencies. If this information is released in an unauthorized manner or prematurely, it could compromise the operations of the agency or if a nation, its security.

To mitigate such disclosures, personnel undergo an investigation that leads to a *security clearance*. This type of investigation delves into the employee's charter, trustworthiness, and loyalty in order to ensure that he or she can be relied upon to hold secrets secure. The theory is that a person who has been found unreliable in one or more of these areas may be a risk when it comes to guarding information. Or the person may be subject to blackmail and forced to surrender information to avoid disclosure of compromising details of past or present activities.

These investigations, therefore, probe personal character; they examine how people conduct themselves in public and private, test their honesty, their financial position, and inquire about any criminal involvement. They also try to gain an appreciation of their emotional and mental strength as well as other issues that might jeopardize their work with important secrets. Foundational to almost all these investigations are checks of national police records and credit worthiness. In addition, most investigations conduct interviews with individuals who know the applicant's personal behavior patterns and ethical temperament.

Countries may have slightly different security classifications, but generally speaking, national security classifications are divided into categories that comprise the following:

Unclassified. Documents that do not require the protection of a security classification. They may include documents suitable for public release and could include those deemed "for official use only."

For Official Use Only. Despite appearing to be a security classification, it is not. The annotation merely alerts the user to the fact that it is protected under a privacy act because it is sensitive data.

Confidential. Information, if disclosed, could reasonably be expected to cause *damage* to national security.

Secret. Information, if disclosed, could reasonably be expected to cause *serious damage* to national security.

Top Secret. Information, if disclosed, could reasonably be expected to cause *grave damage* to national security.

What necessitates a security clearance? In most cases it is based on one's access to classified information. Say, for instance, if an analyst's job involves accessing confidential information, that person would require a security clearance to the level of confidential. If the person's job required him or her to have access to secret information, they would have to obtain a clearance to secret level.

The screening process for personnel usually starts at the application stage with the applicant completing a detailed *personal history statement* (PHS). This is carried out to prevent both the hiring of unethical people, who may in time disclose confidential information, but also to frustrate any attempt at penetration by a hostile agent. In addition to the applicant's full name, current residential address, and date and place of birth, the PHS can include such sub-histories as:

- Marital history;
- Residential history;
- Citizenship history (usually 10 years for a clearance to secret level and 15 years for top secret);
- Educational history;
- Employment history;

- Overseas travel history;
- Military history;
- Criminal history; and
- Details of organizational memberships, as well as character, professional, and credit references.

When reviewing the applicant's PHS, a background investigation will look for any inconsistencies, discrepancies, or unaccountable periods (usually around three months or more). Even though applicants may pass this initial screening process, once they are hired, a probationary period is usually set as a contingency for their possible dismissal should they be suspected of having become a security risk. Similarly, when analysts are promoted or assigned to more sensitive research projects (e.g., top secret), another screening procedure can be conducted, covering the time elapsed from their initial hiring to the present. This is to ascertain if any factors in the analysts' recent past could jeopardize the confidentiality of the information they will be handling.

Another means of safeguarding sensitive information is by drawing up nondisclosure or secrecy agreements. These agreements are intended to create a psychological impression on analysts, reinforcing the importance of protecting information to which they have been entrusted. These agreements are in effect legal contracts and can be used as evidence in a court of law should the analyst be found to be in violation. Nondisclosure agreements are also used for temporary research support staff such as data entry operators, computer programmers, and the like.

STORING, PROTECTING, AND DISPOSING OF INTELLIGENCE DATA

Document Storage

An intelligence unit's first line of defense against penetration by a hostile agent is its external barriers—its walls, doors, and windows. Its second line of defense is the "containers" that house its confidential documents, for example, computer servers, desktop workstations, and filing cabinets/storage shelves. It is, therefore, essential that an intelligence unit identify all documents and records that may be the target of a hostile agent and secure these in containers that minimize the risk of their unauthorized acquisition. The concern is with both the theft of the documents and the undetected theft of the information they contain. So, to further reduce the risk of attack on containers designated for sensitive material, the intelligence unit should not store valuables, such as cash, securities, jewels, and precious metals, in them.

Photocopying sensitive documents or digitally photographing them surreptitiously are the two most likely ways a hostile agent could obtain information without a unit's knowledge. Another method, although difficult to attempt, is to remove the document(s) from the unit's offices, copy them, and then return them to their storage container undetected. To guard against the former case, access codes should be installed on the office photocopier to strengthen this potentially weak security link.

To counteract both possibilities, intelligence units should ensure that containers housing their documents have locks and that the locks are used faithfully. Metal containers, such as filing cabinets, with padlocks offer a reasonably high level of security; however, safes and cabinets with combination locks incorporated as part of their physical structure offer a much higher level of protection.

Secure storage is equally applicable to computer disks, portable and flash USB drives, and tape backups (including offsite storage of backups). The control of keys for an intelligence unit's document containers should follow those guidelines outlined previously.

Document Reproduction

The reproduction of classified documents bearing confidential, secret, and top secret (or in a private intelligence organization—grades I, II, and III) are often marked with the classification of the original material. Only sufficient copies necessary to meet operational requirements are duplicated, and all reproductions are destroyed as soon as they have served their purpose (e.g., concluding a briefing for a decision-making committee). Also, when photocopying sensitive documents, analysts should be cognizant to collect the original(s) before leaving the machine.

Document Safeguards during Use

When confidential documents are not held in secure containers, the analyst using the documents is usually required to:

- Keep the documents under constant visual surveillance;
- Place the documents in a storage container and cover it or turn it face down when an unauthorized person is present;
- Return the documents to their designated storage container after use; and
- In the case of plans, graphs, charts, or other forms of visual aids, they should be labeled with a code name or code number and not openly bear a designation that could identify the project to an unauthorized observer.

Document Disposal

An intelligence unit's waste paper basket is an easily accessible source of information for the hostile agent. For instance, it isn't unreasonable to assume that 80 percent or more of the paper generated by a commercial business or professional consulting firm contains information that is confidential to one degree or another. This is information that, if acquired by a competitor, could adversely affect business performance. This information gathering technique is known in the vernacular as "dumpster diving" and is carried out simply by collecting the day's paper waste before the disposal truck arrives.

An easily overlooked source of information leakage is the photocopier. Spoiled and overrun copies should not be indiscriminately dropped into the waste paper basket. An important piece of equipment for all intelligence units is a document shredder. These devices are so common now that even retail stores carry them as standard items. An alternative for large units is to use a bulk document destruction service. These companies are usually listed in the Yellow Pages.

Another often overlooked source of information leakage is the impressions left on writing pads. To guard against this, a thin piece of acrylic, plastic, or aluminum should be used under the top sheet of all memo pads and writing tablets to prevent impression marks being left on the pad. Stenographic notes, worksheets, sticky notes, and similar items should be destroyed—not just disposed. Needless to say, a readable copy can be obtained from any of these sources, and therefore, they are as dangerous as the originals in the hands of a hostile agent. Dictation recorded on tape or digitally should be deleted immediately after being transcribed.

Computer Workstations

Desktop computers pose specific security problems. The chief risks are from unauthorized hardware and software access and software sabotage. In the main, the best countermeasures are those of sound physical and personnel security and software management. Countermeasures designed to protect computer software and data include:

1. Using passwords to authenticate legitimate users of the system. Passwords should be impossible to guess, so it is advisable to avoid using any name that is common or familiar in the work environment, business, or to the project. Also, names that are meaningful to the user, for example a spouse, child, or pet's name, should be avoided. Passwords consisting of a combination of letters and numbers/symbols are ideal

for a very high level of security. Ensure that the software security program that controls user access does not display the password on the screen when logging on or appears on any printouts. Users should commit passwords to memory; they should never be posted on terminals, work stations, or notice boards. Above all, users should never tell anyone without proper authority a system's password. A system's password should not be changed at regular intervals but randomly to foil any attempt to anticipate security changes.

2. Isolating information of various grades on separate drives and labeling each with its level of sensitivity, that is, unclassified, official use only, confidential, secret, top secret (or grades I, II, III, IV) as outlined in the section on classification of information.

3. Avoiding the use of fixed hard disks for storing sensitive data primarily because they cannot be readily removed for safe storage. A removable/portable hard disk offers a much higher level of security. If a fixed hard drive must be used for work associated with projects involving top secret (grade I) information, the alternative is to store the data on portable or flash USB drives.

4. Storing all disks (data disks or flash drives, master program disks, and backup program disks) in secure containers.

5. Degaussing damaged or defective drives that contain business information before returning them to the manufacturer or retailer for credit.

6. Overwriting drives before using them for information of a lower classification.

7. Disposing printouts as outlined in the section on document disposal.

8. Disposing of portable and flash USB drives by physically destroying them.

9. Shutting down idle terminals.

Countermeasures designed to protect computer hardware include:

1. Bolting the computer base unit to the work station.

2. Locking the server room when technicians are not in attendance.

3. Positioning computer screens to prevent viewing from windows, doorways, or through glass partitions.

4. Allowing only trusted and qualified technical personnel to service or make modifications to a system.

5. Conducting electronic countermeasure sweeps at irregular intervals for bugs or wiretaps.

6. Shielding cables leaving the server room in metal conduit to prevent electromagnetic radiation, which could be intercepted, and to deter illegal tapping.
7. When disposing of old hard drives, use a commercial disk cleaning software package that writes zeros over the entire disk surface. This will leave the disk useable but sensitive data unrecoverable. If the object is to destroy the disk, then after using a software cleanser, drill four holes (e.g., at 12 o'clock, 3 o'clock, 6 o'clock, and 9 o'clock) through the unit so that each punctures the platter.

And countermeasures designed to guard against software sabotage include:

1. Using only commercial software from recognized, reputable software manufacturers or a custom developer.
2. Loading programs from the manufacturer/designer's original copy or downloading directly from the developer's website.
3. Using only programs on an "approved programs list" in order to reduce the possibility of contracting program viruses from non-commercial software.

   If non-commercial software must be used, for example public-domain programs, shareware, freeware, and programs downloaded from the Internet, the software should first be thoroughly examined and tested for the possible presence of the sabotage mentioned above.

   Before screening new, non-commercial, software products for viruses and the like, access to the system's hard disk should be temporarily blocked in order to avoid infection should there be any form of contamination in the program.

   Once new non-commercial software has passed rigorous screening, system users should then be supplied with approved (tested) copies of the program.
4. Implementing or upgrading backup and recovery procedures, which will facilitate a quick and complete reconstruction of a system's programs and data in the event that a saboteur strikes.

Finally, the security measures that an intelligence unit adopts to protect its computer work stations and IT systems should not be discussed with anyone outside of the organization. It is acceptable, however, to acknowledge that measures to combat espionage and sabotage are in place, but the specific techniques and procedures should never be confirmed.

FINAL CONSIDERATIONS

Confidential information should always be regarded as having a finite lifespan. It must be realized that despite the best engineered security plans and the installation of the most sophisticated countermeasures equipment, eventually information that is being guarded will become known to others. The American nuclear fighting capability became known to the entire world on August 6, 1945—the day before it was classified top secret.

Obviously, the best way to keep secrets is to store them in one's head and not communicate them to anyone. This is not a very realistic countermeasure in a business environment. The point to be made, however, is that as more people know about some particular secret and the more that is written and recorded about it, the more likely that the secret will become prematurely known to unauthorized people (either inadvertently or by design).

This was the case with the early American atomic bomb research; a Soviet espionage operation was able to penetrate the top secret project and acquire that information well before the rest of the world knew it existed. The second point to be made is that once intelligence agents know or even suspect that someone is guarding secrets, half of their work is done; their next step is to devise a method to acquire it. The question every military, government agency, and business must consider is: how long is this information going to be secure?

KEY WORDS AND PHRASES

The key words and phrases associated with this chapter are:

- Classification;
- Compromised information;
- Confidential;
- Countermeasures;
- Document sensitivity;
- Hostile agent;
- Information security;
- Official use only;
- Personal history statement;
- Premature disclosure;
- Screening;
- Secret;
- Security clearance;
- Threat level;

- Top secret; and
- Unclassified.

STUDY QUESTIONS

1. List at least six types of information that would be considered sensitive enough to carry a security classification.
2. List the different levels of security classifications, and describe the differences between each.
3. Brainstorm four different categories of hostile agents that your agency might have concerns about.
4. Describe the methods an intelligence analyst should use when disposing of classified material. Does your agency adhere to such methods? If not, how can this aspect of information security be improved?

LEARNING ACTIVITY

What does premature disclosure mean? Does this means that *every* classified document will be released one day? Research the laws in your jurisdiction as they apply to different types of intelligence documents. Determine which ones are subject to a permanent classification—never to be released—and those that are subject to a time limitation, meaning that one day they will be released to the public.

NOTES

1. U.S. Department of Defense, *DoD 5220.22-M: National Industrial Security Program Operating Manual* (Washington DC: GPO, February 28, 2006).

2. U.S. Department of Defense, *DoD 5200.1-PH-1: A Guide to Marking Classified Documents* (Washington DC: GPO, May 2000).

3. Australian Bureau of Criminal Intelligence, *Document Security* (Canberra, Australia: Australian Bureau of Criminal Intelligence, 1987).

4. If sensitive information has been compromised or just "lost," the following guidelines will assist in minimizing the damage that may result:
   - Attempt to regain custody of the documents/material;
   - Assess the information that has been compromised (or subjected to compromise) to ascertain the potential damage, and institute action necessary to minimize the effects of such damage;
   - Investigate to establish the weakness in the security arrangements that caused or permitted the compromise, and alter these arrangements in order to prevent any recurrence; and
   - Take actions appropriate to either educate/counsel/discipline the person(s) responsible.

Appendix

Critical Values of Chi-Square Distribution

Degrees of Freedom	$P = 0.05$	$P = 0.01$
1	3.84	6.64
2	5.99	9.21
3	7.82	11.35
4	9.49	13.28
5	11.07	15.09
6	12.59	16.81
7	14.07	18.48
8	15.51	20.09
9	16.92	21.67
10	18.31	23.21
11	19.68	24.73
12	21.03	26.22
13	22.36	27.69
14	23.69	29.14
15	25.00	30.58
16	26.30	32.00
17	27.59	33.41
18	28.87	34.81
19	30.14	36.19
20	31.41	37.57
21	32.67	38.93
22	33.92	40.29
23	35.17	41.64
24	36.42	42.98
25	37.65	44.31

Note: Tables of critical values of chi-square are available in the public domain via the Internet. The above data were extracted from one such table and adopted for indicative use here.

About the Author

Dr. Hank Prunckun, BS, MSocSc, PhD, is adjunct associate professor of criminal intelligence at the Australian Graduate School of Policing, Charles Sturt University, Sydney. He specializes in the study of transnational crime—espionage, terrorism, drugs and arms trafficking, as well as cyber crime. He is the author of numerous reviews, articles, chapters, and books, including: *Shadow of Death: An Analytic Bibliography on Political Violence, Terrorism, and Low-Intensity Conflict* (Scarecrow Press, 1995); *Special Access Required: A Practitioner's Guide to Law Enforcement Intelligence Literature* (Scarecrow Press, 1990); and *Information Security: A Practical Handbook on Business Counterintelligence* (Charles C Thomas, 1989). Dr. Prunckun has won two literature awards and a professional service award from the International Association of Law Enforcement Intelligence Analysts. He has served in a number of strategic research and tactical intelligence capacities within the criminal justice system over the last twenty-eight years, including almost five years as a senior counterterrorism policy analyst. Dr. Prunckun is also a licensed private investigator and a radio engineer.

Teaching intelligence research and analysis in the Middle East, August 2009. The author (standing, third from left) and an academic [...]